The INDUSTRIAL REVOLUTION

2nd EDITION

by Dave Thompson

ISBN # 0-96361-93-0-6

CLEOPATRA 8726 S. Sepulveda, Ste D-82
Los Angeles, CA 90045 USA

****

ACKNOWLEDGEMENTS
Compiling any kind of encyclopedia requires the hard work and patience of a lot of people, and this one is no exception. Immediate thanks, then, are due to Jo-Ann Greene and Lisa Ridley, both of whom gracefully allowed me to excerpt interviews they conducted with members of Pigface and the Young Gods; to Shawn Scallen, Don Gordon, Brian Robin and Grant Alden for photos; to Rob Cherry, Joe Banks and Jason Pettigrew, Matt Green and Susan King, Chase from Reconstriction, editors at Alternative Press magazine, for conversation and advice; to Art Chantry, ditto; and to all the record companies, press officers, managers and band members who opened their files and searched their memories every time I called with another question. Most of all, though, thanks to Brian Perera, whose original vision this book was, and to my wife, Jo, for typing, proof-reading, and the immortal rejoinder: "I don't care how many times you put King Crimson in there, I'm just going to take them out again." At the time of writing, I still don't know who had the last word.

****

CLEOPATRA
8726 S. Sepulveda, Ste. D-82, Los Angeles, CA 90045 USA

PHOTO CREDITS:

Robert Calvert (p. 11): Brian Tawn
Clock DVA (p. 18): courtesy Contempo Records
Code Industry (p. 19): courtesy Caroline
Coil (p. 20): courtesy Wax Trax
Consolidated (p. 22): courtesy of Nettwerk
Cop Shoot Cop (p. 23-24): Shawn Scallen
Die Warzau (p. 30): Shawn Scallen
Einsturzende / Blixa Bargeld (p. 31) & Live (p. 32): Shawn Scallen; Blixa (p. 32): Fred Berger
Brian Eno (p. 33): courtesy of E.G. Records
Foetus (p. 37): Fred Berger
Robert Fripp (p. 40): Brian Shanley
Front 242 (p. 42): Brian Shanley
Front Line Assembly (p. 43): Third Mind Records
Bauhaus (p. 45): courtesy Beggar's Banquet
Alien Sex Fiend (p. 45): Shawn Scallen
New Order (p. 51): Shawn Scallen
Klute (p.54): courtesy of Zoth Ommog
KMFDM / live (p. 55): Shawn Scallen; group (p. 55): courtesy of Wax Trax
Laibach (p. 61): courtesy of Wax Trax
Legendary Pink Dots (p. 64): courtesy of Play It Again Sam
Meat Beat Manifesto / live (p. 67): Shawn Scallen
Ministry (p. 69-72, 74, 76): Brian Shanley; Al Jourgensen / Brian Oglivie (p. 75, 76): Shawn Scallen
Special Affect (p. 70): Brian Shanley
Revolting Cocks (p. 73, 75): Brian Shanley
My Life With The Thrill Kill Kult / live (p. 78): Paul Valerio
Pankow (p. 87): courtesy of Contempo
Pigface / Chris Connelly (p. 88): courtesy of Wax Trax; Chris Connelly, group (p. 89): Shawn Scallen
Premature Ejaculation / Rozz Williams (p. 92): Dave Hermon
Psychic TV / group (p. 96): courtesy of Wax Trax; live (p. 96): Shawn Scallen
Residents (p. 101): Shawn Scallen
Skinny Puppy/ live (p. 107): Shawn Scallen
Test Dept. (p. 118): Shawn Scallen
Z'ev (p. 132): Don Lewis
BACK COVER–Skinny Puppy / Ogre: Shawn Scallen
****

INTRODUCTION

Much as the authors would like to think it could be, this book is not the ultimate Industrial handbook. In a musical movement which has still to reach its peak, in terms of both music and population, new bands are breaking through constantly and the work of other, perhaps defunct, acts is being re-evaluated daily. Our own first draft of this book contained barely more than half the entries now included; it was only as we delved into the huge canon of music and articles which have grown up around the bands we did include, that the full magnitude of the project became apparent.

From the first electronic pioneers to the latest ambient dreamers, through the Avant Garde to the prophetically Progressive, the roots of Industrial Music are dug in deep - so deep that already one or the other of us is regretting the decision to omit **Captain Beefheart** and **Frank Zappa**. And what of those classicists whose work, whilst not falling into even the most loosely defined rock'n'roll schism, has nevertheless had lasting influence - **Steven Reich**, **John Cage**, **Philip Glass** and so forth? Absence from these pages does not necessarily denote absence from the scene.

Such notable omissions notwithstanding, filed alphabetically within these pages can be found (almost) every major mover on the current Industrial scene, a handful of our own "tips for the top", and around the edges, the vast array of talent which, in some way or another, can be considered influential - which was another area of conflict, and one for which neither of us will now apologize. Also absent, but of immeasureable importance nevertheless, are the many compilations which have introduced to the world some of the scene's most enduring practitioners: Crack of a Belgian Whip (KK, 1990), The Cyberflesh Conspiracy (If It Moves 1992), and so on. Again, apologies, but there's only so much room... and a lot to fit in!

Brief examinations of individual periods within the development of the field - the so-called Kraut Rock boom of the early 1970s, the markedly similar Synthipop movement of a decade later - allow passing comment to be made on bands whose actual contributions to the field can neither be wholly recommended nor utterly disqualified.

It should also be pointed out (once again!) that unless we came to serious blows on the subject, no distinctions have been drawn between true Industrial practitioners and more eclectic spontaneous/improvisational acts. As will become repetitively apparent from the text, the concept of "influence" in the Industrial field is so ill-defined that for every artist who dreamed of Einsturzende Neubaten hitting anvils with hammers, there was another who wanted to emulate Van Der Graaf Generator and crash ships into rocks.

To these, and everybody else who is not content with simply strumming guitars and pounding the drums, this book is dedicated.

"Yeah, Industrial Music. Isn't that what they play in factories to stop the workers going to sleep?" - **Cleghorne Thunk 1990**

****

CHRONOLOGY

1928: The first electronic keyboards, **Leo Theremin**'s *theremin* and **Maurice Martenot**'s *martenot*, are introduced.

1930: They are followed by the Trautonium and the Emicon.

1933: Leo Taubmann's Electronde is unveiled.

1934: Nicholas Obukov's Obukov appears.

1938: Rene Bertrand's Dynaphone completes the first wave of primitve oscillators, all ostensibly under the control of the instrument's operator.

1953: Karlheinz Stockhausen begins working at an electronic music studio near Cologne.

1955: Stockhausen and **Herbert Eimert** commence publication of the periodical Die Reihe, described as the first industrial fanzine.

1957: Stockhausen commences a series of compositional courses at Darmstadt. Future **Can** member **Holger Czukay** is an immediate disciple, and the pair embark on an intermittent writing partnership.

1962: John Cale, Terry Riley, Irmin Schmidt all begin working (separately) with **LaMonte Young.**

1965: Holger Czukay and **Irmin Schmidt** meet; so do **John Cale** and **Lou Reed,** who form **Velvet Underground. The Residents** possibly form. **Robert Moog** invents the Moog Synthesizer.

1966: Mothers of Invention release Freak Out!, arguably the first rock'n'roll album to fully embrace avant-garde ("Return of the Son of Monster Magnet").

1967: Tangerine Dream form in Germany.

1968: Czukay/Schmidt form **Can**; elsewhere, **Amon Duul/Amon Dull II** come together. **Velvets** release White Light White Heat; Cale quits immediately after. **Barrett-**less **Pink Floyd** release Saucerful of Secrets. **Captain Beefheart** signs with **Frank Zappa.**

1969: King Crimson and **Van Der Graaf Generator** debut in England; **Genesis P Orridge** and **Cosey Fanni Tutti** meet at Hull University; in Germany, **Ralf Hutter** and **Florian Schneider** meet at **Terry Riley** exhibition in Berlin. In America, **Beefheart** releases Trout Mask Replica.

1970: The Residents produce their first tapes, and meet **Snakefinger.**

1971: Terry Riley releases Rainbow in Curved Air and collaborates with **John Cale** on Church Of Anthrax. **Eno** joins newly formed **Roxy Music** and debuts on seminal John Peel Session. In Germany, **Kraftwerk** and **Faust** release debut albums, **Neu** form. **Robert Moog** introduces the portable Mini-Moog. Less than 15 years earlier, Columbia Studios had emptied out an entire room to accomodate a machine that fulfilled much the same functions.

1972: Hawkwind's "Silver Machine" makes #2 in UK; **Slapp Happy** record debut album in Germany with **Faust. King Crimson** release live Earthbound LP.

1973: Can release Future Days; **Faust**'s budget-priced Tapes follows. **Henry Cow** release debut album. **Eno** quits **Roxy** and records No Pussyfooting with **Robert Fripp.**

1974: First **Cabaret Voltaire** recordings made. **Devo** and **the Tiller Boys** both form. **Kraftwerk**'s "Autobahn" becomes major UK/US hit single; the album follows.

1975: Henry Cow and **Slapp Happy** collaborate for the first time. **Throbbing Gristle** form. **Boyd Rice** begins recording career. **Tangerine Dream** chart in UK with Rubycon and **Lou Reed** releases Metal Machine Music.

1976: Eno produces first **Ultravox** album. **Residents** release "Satisfaction" 45. **Suicide** debut in New York. In London, **Throbbing Gristle**'s Prostitute presentation is banned, and **Can** have a Christmas hit single.

1977: David Bowie and **Eno** release Low and Heroes, and collaborate on **Iggy Pop**'s The Idiot. In concert with **Kraftwerk**'s Trans-Europe Express and **Donna Summer**'s "I Feel Love", they push electronic music to new commercial heights. **Chrome, Wire, Joy Division** and **DAF** all start up.

1978: Debut releases from **Cabaret Voltaire, Public Image, Human League** and **The Normal. SPK, Birthday Party** and **Foetus** reach UK from Australia, **Suicide** tour Europe to general rioting. **Controlled Bleeding** form.

1979: **Nurse with Wound** and **Einsturzende Neubaten** form. **John Foxx** quits **Ultravox**. **Robert Fripp** releases Exposure; **Bauhaus**' "Bela Lugosi's Dead" launches Goth scene; **Gary Numan**'s "Are Friends Electric" hits UK #1 and gets synthipop moving too. **Public Image** release Metal Box. **Robert Rental** and **Thomas Leer** make debuts.

1980: Clock DVA, Soft Cell, Depeche Mode and **Laibach** form. **Wire** split. **Ian Curtis** dies.

1981: Soft Cell and **Kraftwerk** both hit #1 in Britain. **Ministry, Savage Republic, CTI** and **Front 242** all under starters'orders. **Throbbing Gristle** split.

1982: Christian Death and **Psychic TV** form. **Chris Carter** pronounces Industrial Music dead. Goth wishes it was.

1983: **Laibach** undertake Occupied Europe tour. **Einsturzende Neubaten** release Drawings of Patient OT, **John Cale** creates Music For A New Society, **Depeche Mode** announce Construction Time Again.

1984: Robert Calvert's Freq, **Holger Czukay/David Sylvian**'s Brilliant Trees released. **Ministry** shrug off synthipop trappings; debuts from **Test Department, Skinny Puppy** and **KMFDM**

1985: More debuts, from **Nitzer Ebb, Xymox**, and **Young Gods**

1986: Frontline Assembly form; **Can** reform (why?); CTI label debuts; **Executive Slacks** depart. **Psychic TV** hit the Guinness Book of Records

1987: Snakefinger dies; **Skinny Puppy** release Cleanse, Fold and Manipulate

1988: **Timelords/KLF** hit #1 in Britain. Debuts from **Lard, Meat Beat Manifesto, Renegade Soundwave**. **Laibach** goes disco and **Robert Calvert** goes to sleep.

1989: Debuts from **Consolidated, Hilt, Nine Inch Nails** and **Ministry** spin-off **Pigface**. **Einsturzende Neubaten** split,**Ministry** claim The Mind Is A Terrible Thing To Taste and **Test Department** record with **Brith Gof**.

1990: Depeche Mode's Personal Jesus EP becomes best-selling CD5 in WB's history. Metal Machine Music makes it onto CD.

1991: Wire become **Wir**; **Kraftwerk** release The Mix and **Xymox** put out Phoenix, **Nitzer Ebb** hit with As Is; **Ministry**'s Jesus Built My Hot Rod becomes third highest selling CD5 in WB's history.

1992: Genesis P-Orridge kicked out of UK; **Thrill Kill Kult** hit silver screen in Cool World; **Skinny Puppy** chart with Last Rites and **Nine Inch Nails** with "Head Like A Hole". **Ministry** co-headline Lollapalooza '92.

1993: Ministry tour South America; sporting new tatoos and facial hair, **Depeche Mode** release the long-overdue follow-up to Violator, Songs Of Faith And Devotion; **Nine Inch Nails** hit with Broken; Ex-Hawkwind saxophonist **Nik Turner** links with **Helios Creed** on Sphynx album. Jeff Ward (drummer from NIN, Rev Co and Lard), going through heavy depression, takes his own life.

1994 : Psychic TV Singles and 12" mixes finally get to see the light day on CD.
Genesis P-Orridge joins Pigface for studio album.

DISCOGRAPHIES: Except where noted, discographies are complete to the time of writing. Generally, original releases only are noted. Important reissues are noted in the artist entry.

ACID HORSE see Ministry

A HEAD see X MARKS THE PEDWALK

AJAX

Combining organic-sounding percussion with agoraphobic production techniques, New Yorkers' Ajax debuted on Chicago's Wax Trax label in 1989 with the Mind The Gap EP, a broiling mass of funky dance rhythms over which singer **Mitchell** unleashes a barrage of yelps and shouts. Alternative Press magazine best summed it up, claiming that "[while] many House songs get maddeningly repetitious after a few minutes (seconds?), Mind The Gap has enough diversity to keep everybody interested."

This description also fits the One World EP, released the following year; sadly it also fired accusations that Ajax were all but useless outside of the dance clubs which spawned them. However, an eponymous debut album later in the year proved that Ajax have considerably more at their disposal than they had hitherto let on.

Selected discography: Mind The Gap EP (Wax Trax 1989); One World EP (Wax Trax 1990); Ajax (Wax Trax 1990)

MARC ALMOND see Coil, Foetus, Soft Cell

AMON DÜÜL II

As the logical extension of **Pink Floyd**'s earliest (post-**Syd Barrett**) ramblings, German band Amon Düül II was the first of the so-called Kraut rockers to find a niche in English-speaking hearts.

Formed in a Munich commune in 1968, the original **Amon Düül** splintered very early on into two bands, the more political of the two retaining the original name, the other - comprising **Renate Knaup-Kroetenschwarz** (vocals), **John Weinzierl** (guitar), **Falk Rogner** (keyboards), **Peter Leopold** (drums), **Dave Anderson** (bass) and **Shrat** (bongos) - appending their name with the numeral.

Amon Düül swiftly ceased to be: II, however, signed to UA and in 1969 released Phallus Dei, a gentle collection of space farts and blips set to hypnotic drones and whooshes. Cited by their record company as prime exponents of "space rock", a term which would later be most accurately applied to labelmates **Hawkwind**, Amon Duul II were described as playing "...music which has been left in old musical patterns..., cosmic music [which] liberates the listener in his own fantasia and gives him the liberty of a voyager through time."

The double albums Yeti and Dance of the Lemmings followed, with the band (now settling to a fairly constant line up of Renate, Weinzierl, Rogner, **Lothar Meid** (bass), **Kris Karrer** (Guitar, sax, violin), and **Danny Fischelscher** (drums), moved into less easily categorized areas: Lemmings in particular is a fabulous album, whose effect was summed up by Greg Shaw in Phonograph magazine thus: "as spooky as any music has ever sounded, evoking swirling gusts of icy snow and the emergence of an ancient Frankenstein horror from some Transylvanian cave as the echoing voice of a demented Hitler shrieks garbled nonsense into the wind."

Wolf City and Carnival in Babylon, released on either side of Hawkwind's 1972 commercial breakthrough, were the albums which cemented Amon Düül II's reputation; Live in London from the following year presents the amazing spectacle of Teutonic iciness melting before a delirious UK audience.

This success ironically spelled the end of Amon Duul II as a truly innovative force; subsequent albums often merely tread old ground. This is particularly true of the albums released following the band's departure from UA in 1975. Lemmingmania, a compilation of their pre-75 material is highly recommended. **(see Kraut Rock)**

Selected discography: Phallus Dei (Sunset 1969); Yeti (Liberty 1970); Dance of the Lemmings (Liberty 1971); Wolf City (UA 1972); Carnival in Babylon (UA 1972); Live in London (UA 1973); Vive Le Trance (UA 1974); Lemmingmania (UA 1975).

ANTI-GROUP see Clock DVA

ARMAGEDDON DILDOS

Thanks to their 1st release East West, Armageddon Dildos are synonymous for positive and forceful electronic body music. The energy output of **Uwe Kanka**'s & **Dirk Krause**'s live performances could be transformed into their album That's Armageddon. The new 12" Homocidal Maniac and the album of a similar name links up with this direction.

Armageddon Dildos

Selected Discography: East West EP (Zoth Ommog 1991); Nevermind EP (Zoth Ommog 1991); That's Armageddon (Zoth Ommog 1991); Resist EP (Zoth Ommog 1992); Homicidal Maniac (Zoth Ommog 1992); Homicidal Dolls (Zoth Ommog 1993).

ART BEARS see Fred Frith

ART OF NOISE

Anne Dudley, Gary Langan and **JJ Jeczalik** were the first musicians to sign to ZTT, the label founded in 1983 by journalist **Paul Morley** and producer **Trevor Horn**. This pair were also members of the band, Morley's contributions being to pen sleeve notes and title the rough, largely instrumental, sound collages which did duty as songs.

Like ZTT's greatest asset, **Frankie Goes To Hollywood**, Art of Noise existed less as a band than as a marketing strategy. The Into Battle With The Art of Noise EP appeared on both 12" and as a cassette single, while their debut album, (Who's Afraid of) the Art of Noise! (1984), was released with three different sleeves. More insidious was ZTT's penchant for releasing the same single with different b-sides but the same catalog numbers, a process which was later applied to actual songs - "Beatbox Diversion #2" and the LP version of the band's first hit single, "Close (to The Edit)" are identical, as are "Beatbox Diversion #10" and "Beatbox", the b-side of the "Moments In Love" single. However, "Beatbox", on the Into Battle EP (1983) is a completely different song.

Art of Noise (Morley and Horn excepted) left ZTT in 1985 in disgust at the label's marketing techniques. They signed to Chrysalis' China subsidiary (co-founder **Derek Green**, the man who sacked the **Sex Pistols**) and thereafter moved into more mainstream pop and world music fields - a version of **Duane Eddy**'s "Peter Gunn" gave them an interesting UK hit; "Legs" and its revamp, "Legacy" were even more conventional, and in terms of their industrial output, Art of Noise's best moments remain their stint with ZTT.

Selected discography: Into Battle EP (ZTT 1983); (Who's Afraid Of) The Art of Noise (ZTT 1984); In Visible Silence (China 1986); Reworks (China 1986); In No Sense? Nonsense (China 1989); The Best Of (China 1989); The Ambient Collection (China 1990).

MARTIN ATKINS see Pigface

ATV see Coil, Psychic TV, Punk Rock

BABYLAND

Los Angeles' Babyland, a duo comprising **Smith** (power tools, metal percussion and noise) and **Dan Gatto** (vocals, programming, electronics), got together shortly before Christmas 1989. Describing their work as being "conceptually based on the extreme emotional elements of humankind as reflected in our continual evolution and search for meaning within our own existence", it is better approached as a harsh hybrid of mechanical confusion over minimal electronics.

The first Babyland releases, a four track EP 1991, and the three song Babyland/Reality EP, sold over 3,000 copies with minimal advertising; these are also collected on You Suck Crap, Babyland's 1992 debut album. The band's CD debut was "Mindfuck", included on the Cyberflesh Conspiracy compilation (If It Moves, 1992)

Recliner's "Nose Dive" a one off collaboration with **Vampire Rodents**, is scheduled for release on the Rivet Head Culture compilation (If It Moves; unreleased at the time of writing). **(see Vampire Rodents)**

Selected discography: 1991 EP (Flip 1991); Babyland/Reality EP (Flip 1991); You Suck Crap (Flip 1992).

DAVE BALL see Cabaret Voltaire, Stephen Mallinder, Soft Cell

BLIXA BARGELD see Einsturzende Neubauten

BATZ WITHOUT FLESH

Formed in Delaware in 1986 by **Claude Willey** (the only permanent member), Batz Without Flesh debuted in1988 with a self titled mini album, released on Willey's own Nailed To Sound label and described at the time as "disturbing industrial dance... characterised by hypnotic rhythms and synthetic explosions punctuated by gut-wrenching vocals and agonized screams and moans."

A Million Bricks offered more of the same. but a tighter vision singled the album out for considerable attention, although the Batz themselves were still to shake of the **Skinny Puppy** comparisons with which they were originally saddled. (That said, the group has been described as "more unsettling than **Trent Reznor** could ever wish to be", which is a start!)

Batz Without Flesh

No Memory, "electronic without being cold", finally saw the Batz gaining the identity earlier releases were lacking; This Liquid, recorded with auxiliary Batz **Uncle Pervy** and **Ock Ruin**, is the band's best yet.

Discography: Batz Without Flesh (NTS 1988); A Million Bricks (NTS 1989); No Memory (NTS 1991); This Liquid (NTS 1992).

BAUHAUS see Goth

CAPTAIN BEEFHEART see Chrome, Pop Group, Punk Rock, Residents, Snakefinger3

JELLO BIAFRA see Lard

BLOK 57

Blok 57 Is the cooperation act of **Dirk Ivens** (dive, former The Klinik) & **Guy Van Mieghem** (Vomito Negro). Their first CD of the same name on Zoth Ommog includes a cover of the Normal's "Warm Leatherette." Blok 57 satisfy with their hard & dark electro sound which proves the skills of these two exceptional artists from Belgium.

BIGOD 20

Since their U.S. club smash "The Bog" and "Carpe Diem" (an Umo Detic release in Europe) in 1991, The Bigod 20 cult factor cannot be denied any longer. The major act with Jallokin & Talla 2XLC distinguishes itself by creating intelligent, elaborate and danceable electronic. After four 12" releases, the first album Steel Works!, with new vocalists **Zip Campisi** & **Jeff Johnert** arrived in late 1992.

BODY SINK see Controlled Bleeding

DAVID BOWIE

Alternately a genius, a fraud and a buffoon, Bowie remains amongst the most influential figures to emerge in the past twenty years, which makes it all the more surprising that no more than half of his album catalog is actually worth the plastic upon which it was pressed.

The Man Who Sold The World, a dense semi-metal album recorded during his sojourn of post-"Space Oddity"/pre-Ziggy obscurity, was in many ways the harbinger of the so-called Berlin trilogy recorded with **Brian Eno** and a passing **Robert Fripp** in 1977: earning as early as 1970 a reputation for mechanoid vocals.

Diamond Dogs, in its apocalyptic imagery, too, pointed several future industrialists in the right direction, but it was with Low, Heroes and, to a lesser degree, Lodger that Bowie truly made his mark, obviating his traditional songwriting strategies and throwing the entire session over to Eno.

Of the three, Low is the starkest, primarily because nobody expected such a creation to emerge; Bowie's label, RCA, was said to be outraged, and only with reluctance did they ship the album. Its subsequent success remains one of Bowie's fondest memories.

Boasting seven short, seemingly incomplete musical vignettes on side one; four longer mood-laden instrumentals on side two, Low arrived in January 1977, at the height of punk. Normally a new Bowie album would have been treated with the contempt otherwise reserved for the likes of **Genesis** and the **Rolling Stones**; instead, it was seized upon as a harbinger of a new electric age, and its impact on the many improvisationalists then operating in comparative obscurity was immeasurable.

Retaining the same song side/instrumental side format as its predecessor, Heroes, later that same year, made even Low sound commercial, with only the startling title track bearing much of a tune. Elsewhere, Bowie's own intentions of recording a record as bleak as the Berlin Wall in whose shadow the chosen studio lay succeeded admirably; even time and familiarity have not wholly dulled its impact.

1977 also saw Bowie resurrect **Iggy Pop** as a viable enterprise-come-living legend: with Eno again in tow, he walked Iggy through The Idiot, an album which even Eno likened to having your head encased in concrete. "Mass Production" and "Dum Dum Boys" rank alongside **Kraftwerk**'s near-simultaneous "Metal On Metal" (from the Trans-Europe Express album) as rock's first mainstream acknowledgement of industrial music's potential, while side two of both Low and Heroes lays much of the groundwork for Eno's own subsequent career.

(L to R): Robert Fripp, David Bowie, Brian Eno

But not Bowie's. After a dreadful live album, Stage, served to remind the world that even the best improvised music cannot be reproduced note for note in concert, Lodger completed the Berlin Trilogy with a conventional and, in places, labored collection, a scenario which Bowie also created with the following year's Scary Monsters - an album, ironically, which is still widely touted as the spark plug from which the New Romantic movement was fired. In the event, the opposite is more likely; **Gary Numan**, who by this time made a more convincing David Bowie than Bowie himself had been in years, was already racking up the chart toppers when Monsters appeared.

Rykodisc's CD reissues of Bowie's back catalog appends bonus cuts to each of the listed albums. Beware, however: "Holy Holy", a 1970 non-album single, is replaced by a 1973 version on MWSTW, while both Low and Heroes contain specious 1991 remixes. **(see also Brian Eno, Robert Fripp, Kraftwerk, Gary Numan, Velvet Underground)**

Selected discography: The Man Who Sold The World (Mercury 1970); Low (RCA 1977); The Idiot (with **Iggy Pop** RCA 1977); Heroes (RCA 1977); Lodger (RCA 1979); Scary Monsters (RCA 1980).

BRIAN BRAIN see Martin Atkins

GLENN BRANCA

Ex-**Theoretical Girls** and **Static**, New York guitarist/composer Branca fuses classical guitar and almost conventional rock'n'roll into a barrage most conveniently (but less than accurately) described as a wall of sound.

Most at home on stage, much of Branca's work remains unrecorded; of his legitimate releases, Lesson #1 (**Philip Glass** meets **Joy Division** and the cacophonous The Ascension, dating from 1980-81, were for many years his most interesting works. However, the appearance in 1992 of the decade-old Symphony #2 (The Peak of the Sacred) introduced Branca to a larger, and less eclectic audience, in part through the presence of **Sonic Youth** guitarists **Thurston Moore** and **Lee Ranaldo**, and percussionst **Z'ev**. The Mallet Guitars they play, incidentally, were created by Branca himself. **(see Z'ev)**

Discography: Lesson #1 (99 - 1980); The Ascension (99 - 1981); Who You Staring At? (with **John Giorno** GPS 1982); Symphony #1 (ROIR - 1983); Symphony #3 (Neutral - 1983); Symphony #6 (Blast First - 1989); Symphony #2 (Atavistic 1992).

BRITISH ELECTRIC FOUNDATION see Human League

BROTHERS AND SISTERS

Toronto based **Tom Third** describes Brothers & Sisters as a collaboration between a vinyl archaeologist and his computer, the end result of which is the frequently startling collision of state-of-the-art digital recording techniques and state-of-the-Ark scratchy records.

Melding a variety of dance rhythms, and drawing upon a floating pool of vocalists (**Sahirah, Alvin Watts, Zulema Clas** and rap group **The Last of an Ancient Breed** have all worked with Third), Brothers and Sisters released a stream of independent cassettes before signing to IRS' Nettwerk subsidiary for the release of 1992's Transcontinental Weekend.

FRIEDER BUTZMANN

West Berlin based experimentalist Butzmann enjoyed a flurry of semi-novelty releases around the early 1980s, including the "Stille Nacht Goes Disco" Christmas single, credited to **Butzmann and Santa** (Marat 1980), and "Music for Toilets (Unveroffentlicht)", included on the second Festival der Guten Taten compilation alongside offerings from **Einsturzende Neubaten**, and **Palais Schaumburg**.

Selected discography: Frieder Butzman (Zensor 1981); Vertrauensman Des Volkes (Zensor 1982).

CABARET VOLTAIRE

"It's strange because it's somewhat flattering to have people copying us and **Throbbing Gristle** and other groups, but the whole point was you didn't want people to copy you. The reason you were doing it that way was because it was individual, because you had something basic you were working on which was your own" - **Stephen Mallinder** 1982

Named for a Dadaist club established in Zurich in 1917, Cabaret Voltaire - **Chris Watson, Richard Kirk** and Mallinder - first began playing together in the early 1970s, working with an oscillator and tape recorders. Their first ever recording, "Baader Meinhof", later appeared on the compilation A Factory Sampler.

Later, through the music department of Sheffield University, they were able to expand their arsenal, and early Cab's performances included one show at Sheffield University and another at the Edinburgh Arts Festival, where the band itself stayed at home and sent a cassette and a film along instead. "Is That Me", the b-side of 1979's "Nag Nag Nag" (Rough Trade) single, was recorded at one of the band's 1975 shows.

Cabaret Voltaire 1974-76 is a 1981 release of the band's earliest work, including several pieces which appeared on their first ever release, Limited Edition, a cassette which was indeed limited - just 25 copies were made.

Caught up in the early flurry of punk-era excitement, the Cabs signed to Rough Trade after the label caught them opening for **Buzzcocks**, and in 1978 the group's first record, the Extended Play EP introduced their vision to the world at large. Though the most important cut remains the vicious "Mussolini Head Kick", a cover of **Lou Reed**'s "Here She Comes Now" remains perhaps the most accessible Cabs track from this entire period. Two more typical Cabaret Voltaire tracks appeared on the Factory Sample EP that same year.

The band's first album, Mix Up appeared in the fall, echoing in its wide open use of tape loops and "found" noises the early experimental albums recorded by **Brian Eno** and **Robert Fripp**. Certainly it had little in common with the ferocity of the recently released "Nag Nag Nag" single.

More seemingly haphazard sounds characterized the "Silent Command" and "Three Mantras" singles (both Rough Trade) which followed, while The Voice of America and the in-concert Live at the YMCA 27-10-79 albums continued moving away from the punk arena at the speed of light.

1981 saw Cabaret Voltaire finally hit their peak, in terms of both quality and quantity; the 3 Crepuscule Tracks EP, a Live at the Lyceum cassette and the Red Mecca album testified to the uncompromising creativity which marked Cabaret Voltaire out, with the "Eddie's Out"/"Walls of Jericho" single perhaps the highlight of their entire Rough Trade period. (Early editions came with a free 7" single, "Jazz the Glass"/"Burnt to the Ground"). The band next undertook some soundtrack work for the BBC, released as Johnny Yesno in 1983, plus production for **Tiller Boys** conspirator **Eric Random** and **23 Skidoo**.

Chris Watson quit in October 1981 to take up a technical post in television (he later resurfaced in **the Hafler Trio**); Kirk and Mallinder opted not to replace him, instead pulling in a succession of outside musicians to fill the breach - Eric Random plays guitar and percussion on 1982's live Benefit for Solidarity album (released under the **Pressure Company** pseudonym); he is also in evidence on 2x45, a double 12" package released in June, 1982. (Hail was another live album, recorded this time in Japan.) **Acid Horse**, which paired Mallinder and Kirk with **Ministry**'s **Al Jourgensen** and **Paul Barker** later in the decade extended this practise.

The Cabs parted company with Rough Trade in early 1983, and after a brace of independent singles - "Fool's Game" and a remix of 1981's "Yashar" - the duo signed to Some Bizzare/Virgin (Caroline in the US), debuting with the "Just Fascination" 12". Perhaps the tightest Cabaret Voltaire release yet, it found the band moving into the same mutant disco-oriented territory which Some Bizzare labelmates **Soft Cell** (whose **Dave Ball** produced the record) were mapping out.

This feeling followed the group into the sessions for The Crackdown album in August, 1983, and older fans felt some relief when the group's next few projects - the retrospective TV Wipeout video, and a second Richard Kirk solo album, Time High Friction- brought a return to earlier values.

However it was only a matter of time before the Cabs blended their new found interest in dance music with their original sensibilities, and the September 1984 12" single "Sensoria" was certainly a step in the right direction. The solidly excellent Micro-phonies album arrived in October.

The pattern of video-single-album followed again in 1985, with the Gasoline In Your Eye film project (and accompanying double 12" soundtrack single Drinking Gasoline) proceded by the critically acclaimed "I Want You" 45, and the band's eleventh album, The Covenant, The Sword and the Arm of the Lord.

Harking back more to their first few albums than the more recent neo-mainstream material the band had been involved in, The Covenant was described by Mallinder as an attempt to retain the duo's "identity. We had to get this out because we didn't want to just slip into duplicating or sophisticating what we had done [before]."

However, the more commercial sounds with which the band had dallied were to remain in play, on the Drain Train single in 1986 and, cementing Cabaret Voltaire's recently inked pact with EMI's Parlophone subsidiary, the gentle "Don't Argue" single. The Code, positively the weakest Cabaret Voltaire album ever, followed, its deficiencies only made more glaring by Rough Trade's release of the Golden Moments CD compilation. Two further Rough Trade era compilations, the singles-only Living Legends and Listen Up With Cabaret Voltaire (a collection of non-album oddities) were released by Mute in 1990, as part of a wholesale reissuing of the band's back catalog.

Another compilation Eight Crepuscule Tracks appeared in February 1988 (Crepescule in Europe; Giant in the US). Bringing together all the band's tracks for the French label, including a cover of **Isaac Hayes**' "Shaft", the album again came in harsh contradiction to Cabaret Voltaire's own next project, working with House music producer **Marshall Jefferson**. This led directly to the preposterous "Hypnotized" single, made available in a staggering eleven different mixes in October, 1989, and in the new year, at least seven mixes of "Keep On". The album Groovy, Laid Back and Nasty maintained the barrage. The band also toured the House circuit.

The group's flirtation with House proved shortlived - as did their career with EMI. Defecting to Les Disques du Crepescule, Body and Soul in 1991 was recorded in a week, and with its attention to unconventional detail, found the band moving back into the musically subversive territory it had traveled at its peak.

"It was back to basics," Mallinder said. "We weren't consciously saying 'Oh, we'd better write a few tracks on here that are really catchy', which you... felt you had to do with the last LP at EMI. We went in with a very different attitude." The 44 minute Colours EP continues in this welcome direction. (**see Haffler Trio, Richard Kirk, Stephen Mallinder, Ministry, Punk Rock, Soft Cell, Tiller Boys**)

John Cale

Discography: Limited Edition (private cassette 1976); Extended Play EP (Rough Trade 1978); Mix Up (Rough Trade 1979); Live at the YMCA 27-10-79 (Rough Trade 1980); The Voice of America (Rough Trade 1980); 3 Crepuscule Tracks EP (Crepescule 1981); Cabaret Voltaire 1974-76 (Industrial 1981); Red Mecca (Rough Trade 1981); Live at the Lyceum (Rough Trade 1981); Benefit for Solidarity (Pressure Company Paradox 1982); 2x45 (Rough Trade 1982); Hai! (Rough Trade 1982); Johnny Yesno (Doublevision 1983); The Crackdown (Some Bizzare 1983); Micro-phonies (Some Bizzare 1984); The Covenant, The Sword and the Arm of the Lord (Caroline 1985); Drain Train EP (Doublevision 1986); The Code (EMI Manhattan 1987); Golden Moments (Rough Trade 1987); Groovy, Laid Back and Nasty (Parlophone 1990); Body and Soul (1991); Colours EP (Mute 1991).

JOHN CALE

Although Welsh-born Cale himself has always insisted he has nothing to do with rock'n'roll, there's no denying that his contribution to it is enormous. So enormous, in fact, that even when he distances himself totally from it, he remains within a framework which can be very easily understood.

Work with **Aaron Copeland** was followed by a stint with **LaMonte Young's Theatre of Eternal Music**, allegedly captured on 1962's limited edition album Dream Syndicate. Then came two years with the seminal **Velvet Underground**; upon departing in 1968, ostensibly following constant arguments with **Lou Reed** over the band's true direction, Cale busied himself with production work (**Nico, the Stooges**) and a very fluffy pop album Vintage Violence, before linking with avant-garde master **Terry Riley** (another former member of Lamonte Young's Dream Syndicate) for The Church of Anthrax.

A dense, and in part unlistenable album, the record was wholly improvised, Cale working out the basics with **Bobby Columby** (of **Blood Sweat and Tears**) and **Bobby Gregg** before Riley arrived to overdub "a whole mess of stuff". He then walked out, leaving Cale to mix the proceedings; returned, heard the finished thing and walked out again. The listening public felt much the same way.

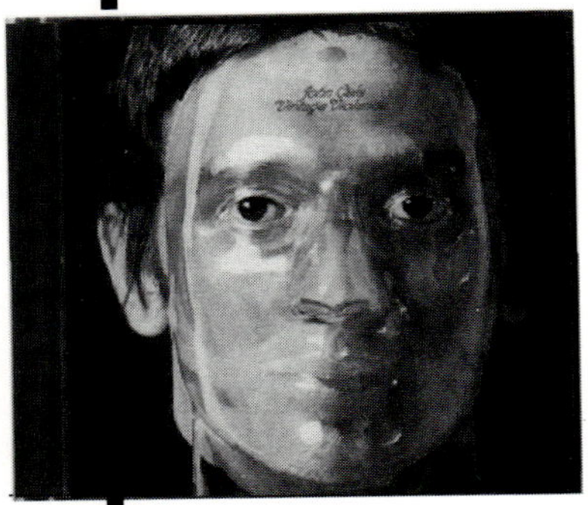

"After [that]," Cale continues, "I figured that I wouldn't mess around with rock'n'roll any more." He recorded Academy in Peril, an uneasy three-way meeting of classical, avant-garde and - yes - rock sensibilities, that didn't really meet the demands of any category. Another noble failure which Cale now describes as "Academy in Pearl Harbor", it marked the end of his most experimental phase, and, for the remainder of the 1970s, Cale concentrated on kicking out gut-busting rock'n'roll records interspersed with some melodies of absolute beauty, several in collaboration with **Brian Eno** (an on-off partnership which has been maintained until today).

By 1979, however, he was up to his old tricks again, going on the road with his heaviest band yet and reintroducing US audiences to some of the cacophonic magic that had been the Velvets' stock in trade: the Sabotage and Even Cowgirls Get The Blues albums capture Cale at his dissonant best, pushing conventional instruments to their limits, and using his own voice as a battering ram.

Music for a New Society, Cale's most overtly industrial album, is by nature a balladic creation, its weight being conveyed by the sheer power of suggestion embodied in the songs. The mood breaks only once, when the band kicks in on "Changes Made". "And that shouldn't be there," Cale insists. "The record company forced me to do it." His own vision was to recreate Nico's Marble Index, an album which Cale himself once described as the nearest thing to suicide ever released.

A stormy decade followed, Cale resuming his uneasy alliance with rock'n'roll, but putting further muscle behind his efforts to create classical music of lasting import. Aspects of Words for the Dying and Paris d'Etudes succeed in this, although for the most part this brilliant songwriter/composer seems content to exist in a shadowy world several notches below the assumed masters of the avant-garde. Paris, incidentally, includes a hitherto unreleased early Velvets cut, the brooding sonic battery of "Booker T". (**see Brian Eno, Terry Riley, Velvet Underground, LaMonte Young**)

Selected discography: Church of Anthrax (with **Terry Riley** Columbia 1971); Academy in Peril (Reprise 1972); Sabotage (Spy 1979); Music for a New Society (Ze 1983); Words for the Dying (Opal 1990); Even Cowgirls Get The Blues (ROIR 1991); LaMonte Young III (recorded 1961-63, 50 Skidillion Watts 1991); Paris d'Etudes (Disques du Crepescule 1992).

Robert Calvert

ROBERT CALVERT

South African born Calvert drifted into **Hawkwind** around 1971, as they were recording their epochal In Search of Space album. Performing as resident poet only, a role he would

Hawkwind

Dave Brock

supplement with full vocals later in the decade, Calvert was a familiar sight at Hawkwind gigs, but did not really make his presence felt on vinyl until the stunning Space Ritual Alive album, by which time Hawkwind had already enjoyed their greatest hit, "Silver Machine" (which Calvert co-wrote).

Featuring much of the recently released Doremi Fasol Latido studio album, Space Ritual and the markedly similar Space Ritual Volume Two capture Hawkwind at their peak. (Dating from some months later, Bring Me The Head of Yuri Gagarin is another excellent live set, marred by sub-bootleg sound.)

Hawkwind at this time were walking a thin line between primal heavy metal and the wider expanses of what was then lovingly dubbed Kraut Rock (as exemplified by labelmates **Can**, **Neu** and **Amon Duul II**); amongst the ideas they introduced to their sonic freak-out was an electronics expert whose function (predating **Roxy Music**'s **Eno** by more than a year) was to "treat" the band's own sound: it was **Dik Mik** who also hit on playing records backwards through his equipment while the band was in full flight, and though it is but an insignificant part of the ultimate whole, his "Electronics #1" on Space Ritual is worth a listen.

Also heavily involved at this time was saxophonist **Nik Turner**, later leader of **Sphynx**. That band's late 1970s album, Xitintoday, has lately been rerecorded for release by Cleopatra (as Sphynx, credited to Turner alone) with contributions from **Tommy Grenas** and **Len Del Rio** from **Pressurehed** and **Helios Creed** of **Chrome**.

Calvert divorced himself from Hawkwind after Space Ritual. They went on to record two more excellent albums, highlights of which feature on both Stasis and, with the addition of many non-album cuts, Psychedelic Warlords. Calvert, meanwhile, produced two staggering solo albums, Captain Lockheed and the Starfighters and Lucky Lief & The Longships, the latter produced by Eno. Both find him carving out a niche in the pre-industrial heartland, through the use of sound collages, spoken word passages which are integral to the overall musical concept, and various bizarre treatments - from the latter, "The Storm Chant of the Skraelings" is of particular interest.

Returning to Hawkwind in 1975, for the sequence of albums beginning with Astounding

Sounds and winding up with Live 1979 (the best of which are collected on the Spirit of the Age retrospective) Calvert's work locked into a conventional musical framework until 1985 brought Freq, a mini-album of awesome proportions.

A contemplative study of 200 years of industrialization, Freq opens with "Ned Ludd", thoughts on the armed uprising against machines which shook England's embryonic manufacturing society in the early 1800s, and closes with "Work Song", espousing an ethic closer to Japanese thoughts on the workplace than western ideas. The reality of that latter condition was brought home by the snatches of conversation from the striking miners' picket lines -Freq was recorded and released at the height of Britain's crippling miners dispute. Instrumentation on Freq was in keeping with the album's subject matter; heavy, rich electronics power-pulsing the music into shape. (The Freq CD appends an early 1980s single, "Lord of the Hornets").

Test-tube Conceived, two years later, continues Calvert's fascination with artificial technology; a colder and more distanced album, it was nevertheless a powerful statement.

Calvert died from a heart attack, aged 43, in 1988. Live at the Queen Elizabeth Hall and Blueprints from the Cellar, comprising demos for Calvert's proposed fifth album, are the only posthumous release of note. (**see Chrome, Brian Eno, Punk Rock**).

Selected discography: HAWKWIND (includes non-Calvert albums) In Search of Space (UA 1971); Doremi Fasol Latido (UA 1972); Space Ritual Alive (UA 1973); Hall of the Mountain Grill (UA 1974); Warrior on the Edge of Time (UA 1975); Astounding Sounds Amazing Music (Charisma 1976); Quark, Strangeness & Charm (Charisma 1977); Hawklords (as **Hawklords** Charisma 1978); PXR5 (Charisma 1979); Live 79 (Bronze 1980); Bring Me The Head of Yuri Gagarin (Demi Monde 1983); Space Ritual Volume 2 (APK 1985); Spirit of the Age (Virgin 1988); Stasis (EMI 1990); Psychedelic Warlords (Cleopatra 1992).
ROB CALVERT: Captain Lockheed & The Starfighters (UA 1974); Lucky Lief & The Longships (UA 1975); Hype (1980); Freq (Flicknife 1984); Test-Tube Conceived (Demi Monde 1986); Live at the Queen Elizabeth Hall (Clear 1989); Blueprints from the Cellar (1992).

CAN

"They are the best thing that ever came from the continent" - liner notes to Monster Movie.

Can

Germany's Can remains one of the most influential bands of the pre-industrial age, their input being felt in the work of bands as disparate as the **Fall**, **PiL**, **New Order** and **Cabaret Voltaire**, but drawing in turn from **Stockhausen**, **John Cage** and **LaMonte Young**. Indeed, keyboard player **Irmin Schmidt** studied under Stockhausen, and worked with both Young and **Terry Riley** during a stay in New York. Immediately upon returning to Cologne, he formed **Inner Space** with **Holger Czukay**, guitarist **Michael Karoli**, drummer **Jaki Liebezeit** and American flautist **David Johnson**.

Drawing from the examples of the **Velvet Underground**, **Jimi Hendrix** and **Frank Zappa**, but blending their musical ideas with "found" sounds - the Paris riots, choir music and so on - Can's first recordings were primitive but exciting. They were finally released on cassette in France in 1985 as Prehistoric Future.
The arrival of a second American, **Malcolm Mooney**, precipitated the adoption of the name Can, late in 1968, and in this form the group released its first single, "Kama Sutra".

Johnson quit shortly after, and it was as a four piece that Can recorded both their debut album, Monster Movie, and the out-takes which comprise the Delay 1968 album released in Germany in 1981.

Originally released in a limited edition of 500 copies, Monster Movie was reissued on United Artists in 1970, and instantly parachuted Can into the forefront of the progressive rock consciousness. The release, on UA's Liberty subsidiary, of Soundtracks added to the band's appeal, their ethereal improvisations lending much to the increasingly urgent strivings of the UK underground. Early **Hawkwind** albums are only the most obvious of Can's many debtors - check out the 14 minute "Mother Sky"!

Mooney quit during the recording of Soundtracks; he was replaced by **Kenji Suzuki**, a Japanese busker Czukay and Liebezeit met outside a Munich cafe, and by early 1971 the second official Can album, the double Tago Mago, was ready for release. One track from this album, "Spoon", became Can's first homeland #1, and the band made its first visit to the UK shortly after.

Ege Bamyasi Okraschoten followed, but this stage of the band's existence was rounded off with the release of Can's fifth album, the brilliant Future Days. More than any other record in the group's catalog, Future Days remains the yardstick by which all future (and past) excursions into electronic music were to be measured, featuring as it does such lengthy and perfectly realized visions as "Spray", "Bel Air" (retitled "Spare A Light" in the UK) and the title track.

Suzuki quit to become a Jehovah's Witness shortly after Future Days brought Can the greatest reviews of their career, and with Schmidt and Karoli taking over on vocals, Can toured extensively (and exhaustingly - one show, in Berlin, lasted seven hours!), before going in to record Soon Over Babaluma, as explosive a record as its predecessor.

1974 brought Limited Edition, a stop-gap collection of out-takes dating back to 1969 (expanded to a double album, this became Unlimited Edition in 1976); a more representative sampling of all that Can had accomplished was released the following year, after the group left UA for Virgin: The Classic German Rock Scene culled selections from every Can album to date. An attempt to repeat this album's success, the budget-priced UK collection Opener, is somewhat less indispensable. Can itself, incidentally, was now hard at work on the band's next album, Landed.

Less satisfying than its predecessors, Landed suffered most from Kareli's sudden realization that he was not a natural frontman. Attempts to bring in a vocalist, however, proved unsuccessful (the legendary **"Magic" Michael Cousins** was one of the handful of singers who passed through the band's ranks), and an unchanged line-up recorded Flow Motion, the band's most uncharacteristic, and contrarily successful, album to date. "I Want More", a disco song written by **Pink Floyd** guitarist **Dave Gilmore**, gave the group a UK Top 30 hit in the fall of 1976. (A similarly intentioned follow-up, the hymn "Silent Night", mercifully failed to chart.) A number of unreleased cuts from this period were brought together on a 27 track cassette, Onlyou (Pure Freude) in 1982.

Former **Traffic** bassist **Rosco Gee** joined the band at this point; he was followed by bandmate **Reebop Kwaku Baah**, but Can's flirtation with one of the best bands of the early part of the decade was to cut little ice at its punk-whipped end. Saw Delight in 1977 was very much the new boys' show, with Can's own individuality lost for much of the time. They were dropped by Virgin shortly after. (The Incandescence compilation, dating from 1981, is a fine introduction to this stage of the band's career. Sleeve notes, incidentally, were penned by **Buzzcock Pete Shelley**.)

Can finally broke up following their self-titled 1979 album, with all four members going onto occasional solo careers. A 1986 reunion, with Mooney on board for the first time in 15 years, spawned the somewhat forgettable Rite Time album. **(see Holger Czukay, Kraut Rock, Pink Floyd, Terry Riley, Irmin Schmidt, Tiller Boys, LaMonte Young)**

Discography: Monster Movie (Music Factory 1968); Soundtracks (Liberty 1970); Tago Mago (UA 1972); Ege Bamyasi Okraschoten (UA 1972); Future Days (UA 1973); Limited Edition (UA 1974); Soon Over Babaluma (UA 1974); Landed (Virgin 1975); Unlimited Edition (UA 1976); Flow Motion (Virgin 1976); Opener (Sunset 1976); The Classic German Rock Scene (UA 1976); Saw Delight (Virgin 1977); Out Of Reach (Lightning, 1978); Cannibalism (UA 1978); Can (Laser 1979); Incandescence (Virgin 1981); Delay 1968 (Spoon 1981); Onlyou (Pure Freude 1982); Prehistoric Future (Tago Mago 1985); Rite Time (Mercury 1989); Cannibalism 2/3 (Mute 1990).

CHEM LAB

Having provided support on **Nine Inch Nails**' first headlining US tour, Chem Lab spearhead the first wave of Industrial bands to be operating in the wake of the movement's early 90s' commercial breakthrough.

Chem Lab

Based in Washington DC, the trio, comprising **Jared**, **Dylan Thomas** (presumably no relation) and **Joe Frank**, debuted in 1990 with the Ten Ton Pressure EP, a formative, but promising offering which was well received on both sides of the Atlantic. Industrial Nation has described them as **"NIN** meets **Skinny Puppy** meets 'Twitch' era **Ministry"** - high praise indeed!

Discography: Ten Ton Pressure (5th Column, 1990); Burn Out At The Hydrogen Bar (5th Column 1993).

CHRIS CARTER see Chris & Cosey

NICK CAVE see Foetus, Soft Cell

CHRIS & COSEY

"Sex is...important in our lives and it invariably overlaps into our work. It's often missing in electronic music, so we consciously inject it into what we do." - **Chris Carter** 1989

Having exchanged their own long-time partners for one another shortly before the demise of **Throbbing Gristle**, Chris Carter and **Cosey Fan Tutti** have become the **Sonny and Cher** of industrial music, albeit with a somewhat more lasting vision of what that music means.

Creative Technology Institute, the project which got their new career off the ground, has released only one record under that name, 1984's Elemental 7, the soundtrack to a lengthy CTI video. As Chris and Cosey, however, the pair have a phenomenal catalog of thought-provoking and often utterly scarifying music behind them, of which only their debut, Heartbeat, seems to deliberately tread in Throbbing Gristle's footprints. (Chris also released a solo single around this time, "The Space Between".)

Trance is a darker, more contemplative work, while 1984's Songs of Love and Lust has been compared to a leftist **Depeche Mode** - were Depeche fans of **Nico**. Alongside 1985's Techno Primitiv, this remains the safest of the duo's albums, and listeners would be better advised to hear **Daniel Miller**'s Silicone Teens project before taking either of these discs seriously. Material from this entire period is collected on the compilation Reflection. A second collection, **Conspiracy International**'s Collectiv One was released in Austria only. (There is also an enjoyable live album, dating from 1983: European Rendezvous.)

The pop longings which these records seemed to exhibit were not, however, fruitless. Take Five, released in Canada only, was an EP of dynamic proportions, whose brilliance was only heightened by the inclusion of two remixed oldies. (The Exotica album, sadly, merely repeats, instead of enlarging upon, Take Five's assets, but is a worthy acquisition regardless.)

Frustrated by Rough Trade's insistence on releasing only Chris and Cosey records, the duo formed their own CTI label in 1986, to handle Carter's first solo album, Mondo Beat; Cosey's Time To Tell followed two years later, and both sets are recommended, demonstrating as they do precisely what each party brought to the duo - and how much they need the other beside them. The slight superiority of Cosey's album again brings the Sonny and Cher analogy to mind.

Action in 1987, and two 1989 albums, Allotropy and Chris and Cosey, followed; 1988 also saw the appearance of Core, which united the duo with ex-Throbbing Gristle man **Peter Christopherson**, **Robert Wyatt**, **Boyd Rice** and sundry others, somehow dispenses with its makers' own pasts and aims more for **This Mortal Coil** territory. Credited to **Core**, it is a very easy album to get into, but should not be used as an introduction to Chris and Cosey themselves.

Released in Belgium only, Trust saw the duo reunited on an album which in many ways harked back to the majesty of Trance, or even the latter-day Throbbing Gristle. The title track in particular was mesmeric, and provided the basis for much of 1991's impressive, but occasionally monotonously skeletal Pagan Tango.

The Synaesthesia CD single is more ambient, and includes a **Daniel Miller** remix heavily redolent of Depeche Mode (!). (see **Daniel Miller, Boyd Rice, Throbbing Gristle**)

Discography: Heartbeat (Rough Trade 1981); Trance (Rough Trade 1982); Songs of Love and Lust (Rough Trade 1984); European Rendezvous (Doublevision 1984); Elemental 7 - The Original Soundtrack (as **Creative Technology Institute** Doublevision 1984); Techno Primitiv (Rough Trade 1985); Take Five EP (Nettwerk 1986); Mondo Beat (**Chris Carter** Conspiracy International 1986): Exotica (Nettwerk 1987) Action (Licensed 1987); Time To Tell (**Cosey Fanni Tutti** Cathexis 1988); Core (**Core** Nettwerk 1988); Collectiv One (**Conspiracy International** CTI 1988); Trust (Play It Again Sam 1989); Chris and Cosey (Play It Again Sam 1989); Allotropy (Staaltape 1989); Reflection (Wax Trax 1990); Pagan Tango (Wax Trax 1991); Synaesthesia EP (Wax Trax 1992).

CHRISTIAN DEATH see Goth, Premature Ejaculation

CHROME

Born out of the first generation of San Francisco punk bands, Chrome - **Damon Edge** (synths, vocals) and **Helios Creed** (guitar, vocals) maintained a steady shockwave of inspired pre-industrial psychedelia over the course of nine essential albums, released between 1977-83.

Wearing their influences - **Can, Hendrix, Beefheart** and **the Residents** - loudly on their sleeves, their music was a wild, and totally abandoned sequence of frantic noise, a precursor of the hardcore thrash which eventually engulfed the Bay Area, but retaining sufficient emotional and lyrical distance to prevent the group from ever getting bound up in that scene. **Sonic Youth** are perhaps the most obvious of their latter-day adherants.

Their first two albums, The Visitation (recorded by Edge alone) and Alien Soundtracks, remain required listening for anybody wishing to comprehend the relevance of psychedelia in modern society, at the same time as laying the groundwork for many of the early 1980s synth-dustrial crew: Half Machine Lip Moves, from 1979, and the blistering 3rd From The Sun, on the other hand, show Chrome to be both following, and pre-empting, developments elsewhere - the ghost of **Gary Numan** is particularly welcome on No Humans Allowed, Read Only Memory dallies with the **Suicide** sound, and first time listeners will doubtless pick out a million other reference points as they play through this stunningly original band's catalog.

Most highly recommended of the original releases is Chrome Box, a 1982 set which brought together the band's second, third, sixth and eighth albums, plus the double Chronicles, which is a maddening, but immensely enjoying, flick through Chrome's unreleased back pages. This was later released separately in Germany.

Chrome broke up in 1983 and both Edge and Creed went solo, with less than satisfying results; Edge's attempt to reform Chrome (without Creed!), too, was disappointing, although Eternity, credited to Damon Edge and Chrome, does have a few moments of inspired madness. But from Into The Eyes of the Zombie King on, Chrome's discography is for hard cases and import freaks - most of it was released only in France (Mosquito) and Germany (Dossier) only.

Creed, contrarily, has remained constantly fluid, with The Last Laugh pairing him - to excellent effect - with **Skin Yard**'s **Daniel House**, an indication of just how dense the set was. He was also heavily involved in **Nik Turner**'s Sphynx album. (**see Robert Calvert**)

Discography: The Visitation (Siren 1977); Alien Soundtracks (Siren 1978); Half Machine Lip Moves (Siren 1979); Read Only Memory EP (1979): Red Exposure (Siren 1980); Inworlds (Siren 1981); Blood

On The Moon (Siren 1981); 3rd From The Sun (Siren 1982); No Humans Allowed (Siren 1982); Chrome Box (Subterranean 1982); Raining Milk (Mosquito 1983); Chronicles (Dossier 1984); Into The Eyes of the Zombie King (Mosquito 1984); The Lyon Concert (Dossier 1985); Another World (Dossier 1986); Eternity (Dossier 1986) Dreaming In Sequence (Dossier 1987); Live In Germany (Dossier 1989); Alien Soundtracks II (Dossier 1989).
HELIOS CREED: X Rated Fairytales (Subterranean 1985); Superior Catholic Finger (Subterranean 1989); The Last Laugh (Amphetamine Reptile 1989); Boxing The Clown (A.R. 1990); Lactating Purple (Amphetamine Reptile 1992).
DAMON EDGE: Alliance (New Rose 1985); The Wind Is Talking (New Rose 1985); Grand Visions (New Rose 1986); The Surreal Rock (Dossier 1987).

CLAN OF XYMOX see Xymox

CLAWHAMMER see Devo

Clay People

CLAY PEOPLE

A chaotic electronic release of fire ritualistic beats. Interwoven by subsonic crash, beauty and pain through programmed keys, inorganic steel of live, grunge guitar pushed through the presence of dark melodic vocals.

Discography: Toy Box (Maltese 1991); The Fire Tribe (Re-Constriction 1993).

CLOCK DVA

Straddling the quite disparate fences erected on the one hand by **Throbbing Gristle**, and on the other by Britain's flirtatious early-1980s rediscovery of sixties soul music, Clock DVA emerged from the industrial wasteland of Sheffield, England, with contributions to the 1980 Hicks from the Sticks ("You're Without Sound") and First 15 Minutes ("I'm So Hollow"/"Brigade") compilations. These were followed by the "4 Hours"/ "Sensorium" single and the improvised White Souls in Black Suits, a privately released cassette which blurred the group's own intuitive grasp of soul music with near-constant roar of white noise.

Clock DVA

Less dance-oriented, 1981's Thirst continued **Adi Newton**'s interest in machinery, and overcoming the departure of most of the band (to the decidedly discoid **Box**), Newton reinvented Clock DVA as the logical successor to **Public Image Ltd**'s epochal Metal Box album. Advantage, the group's major label debut (Polydor UK) arrived in 1983; it includes "Breakdown", a minor hit single that same year. (Excerpts from Advantage were released in the US on the Breakdown EP.)

Clock DVA disintegrated following the release of Advantage; Newton kept the faith via **The Anti-Group**, an industrial jazz combo which nevertheless failed to grab the attention it deserved, and in 1989, he reconstituted Clock DVA (with **Robert Baker** and **Dean Dennis**) for The Hacker/The Act, an album which melded early elements of the original band with the more contemporary innovation of sampling.

It was not an altogether satisfying mix, but undeterred, Newton continued on this line for 1990's Buried Dreams, a conceptual piece which reprised the title cuts off the previous album on a record otherwise heavily indebted to **Suicide** - both the group and the concept. Man Amplified, according to Newton, boasts a little more "soul".

Discography: White Souls in Black Suits (Industrial 1980); Thirst (Fetish 1981); Advantage (Polydor 1981); Breakdown EP (Relativity 1983); The Hacker/The Act (Wax Trax 1988); Buried Dreams (Wax Trax 1990); Man Amplified (Contempo 1991).

CLUSTER see Eno, Kraut Rock

CODE INDUSTRY

Code Industry was born out of the ashes of **Code Assault**, the seminal Detroit electro-collective that combined the popular techno sound of their hometown and influences picked up from European bands such as **Front 242**, **A Split Second** and **Meat Beat Manifesto**. **Rob Myers** (keyboards), **Ethan Nep Sevy** (keyboard, programming), **William Keith** (drums) and **Kyl Chris** (keyboards, drums and programming, vocals) carried on as Code Industry, refining and radicalising the initial approach of Code Assault.

Code Industry

Their first album, Method Of Assembly, released on Antler-Subway in 1990 introduced them to the world of industrial music and received considerable critical attention. On the mini-album, Structure, and the recent Suffer CD5, Code Industry further developed their uncompromising musical vision. With a constantly shifting and flexible framework, they combine European beats, acid basslines, house elements and industrial overtones that completely set them apart from other electronically informed bands.

Code Industry are a bit of an enigma in the world of electronic music. Not only are they the only U.S. band signed to the prestigous Antler-Subway label, they are also the only all African-American electronic band. Once described as "the electronic **Living Color**", Code Industry are closer in content to **Public Enemy**. Combining dialogue samples and lyrics,

Young Men Coming To Power is a militantly political album tackling subjects such as racism, the hypocrisy of patriotism and the media.

Select discography: Method Of Assembly (1990 Antler-Subway); Young Men Coming To Power (1992 Caroline).

COIL

John Balance was already recording sporadic private cassettes under the name of Coil when **Peter Christopherson**, on the run from **Psychic TV**, joined him in 1982, and two years later released the How To Destroy Angels EP in Belgium - a one-sided effort comprising 17 minutes of gonging.

Signing to Some Bizzare, a label whose flirtations with the pop mainstream (via **Soft Cell** and **The The**) belied the strength of experimentation going on elsewhere in its catalog, the duo enlisted **Foetus** overlord **Clint Ruin**, **Alex Ferguson** (ex-**ATV**) and **Stephen Thrower** for Scatology, a lengthy, and often despairing percussive symphony over which were layered synthesizers, guitar feedback and assorted samples.

Coil

Around this time Coil played a now-legendary show with **Soft Cell**'s **Marc Almond**. "It was great," Christopherson remembers. "Marc was reading poetry while John gave himself enemas of blood onstage. **Nick Cave** walked out on us saying we were too gay."

The following year, 1985, Coil contributed three tracks - recorded with **Boyd Rice** - to the Belgian Nightmare Culture, an album split between themselves and **Current 93**. But a two year lay-off preceded Coil's next brace of releases, the three track Anal Staircase EP, which lent two of its tracks to the Horse Rotorvator album, an album which dispensed almost wholly with the danceable percussives of Scatology, and moved instead almost into the territory of what was becoming known as "World Music" - African and Middle Eastern motifs abound, as does a real, and very personal, awareness of death. "When we were recording [it]," Christopherson later admitted, "it was the beginning of the AIDS epidemic and our friends were beginning to die." The mood heavily influences the record.

At this point Coil was contacted to supply a soundtrack to **Clive Barker**'s Hellraiser movie; their submission was not used, but Unreleased Themes From Hellraiser nevertheless appeared in 1988, and achieved better reviews than either the movie or its eventual soundtrack could ever have dreamed of. Surplus material from this project, together with out-takes from its two precedessors, incidentally, appears on 1988's 18 track Gold Is The Metal With The Broadest Shoulders collection. Also included were several non-album cuts, lifted from past compilations and one-offs. Unnatural History continued this archaeological exercise, and contains the full How To Destroy Angels EP, plus Coil's contributions to Nightmare Culture.

Regrouping in 1991, Coil released Love's Secret Domain, a self-confessed "party album", which combines their traditional ambient work with a piece-meal study of modern trends

- acid house, hip hop and so on. Amongst a galaxy of featured stars are Marc Almond and **Annie Anxiety**. (**see Current 93, Foetus Inc., Psychic TV, Boyd Rice, Soft Cell**)

Discography: How To Destroy Angels EP (Laylah 1984); Scatology (Some Bizzare 1984); Nightmare Culture (with Current 93-Laylah 1985); The Anal Staircase EP (Some Bizzare 1987); Horse Rotorvator (Some Bizzare 1987); Unreleased Themes From Hellraiser (Solar Lodge); Gold Is The Metal With The Broadest Shoulders (Threshold House 1988); Unnatural History (Threshold House 1990); Love's Secret Domain (Wax Trax 1991).

COMPOUND

One of several guises utilized by **Andrew Szava-Kovats**, a Massachussetts-based electronic composer. His earliest recordings were in the experimental sound collage vein: Kovats first release, in 1983, was **Data-Bank A**'s Spiritus Sanctus, an electronic dance-oriented cassette on Kovat's own KO city Studio label. Dominion, debuting the following year. offers softer, more melodic fare; **Parade of Sinners** and **Compound** (a trio comprising Kovats along with **Ferd** and **Lewis Francis**) are the harshest of Kovat's industrial visions.

Aside from his own music, Kovats has been instrumental in spreading industrial music throughout the US via the four part Grindstone compilations released on K.O. City. Kovats' own bands of course appear on these sets. with **Compound's** recorded debut being included on 1988's Turn Of The Grindstone.

Discography: COMPOUND: Compound (Dossier 1989); Cinema Verite (Dossier 1990)
DATA-BANK A: Spiritus Sanctus (KO City 1983); The Citadel (KO City 1984); Language Barrier (KO City 1985); Access Denied (KO City 1985); Continental Drift (Lively Art 1987). The Birth of Tragedy (Lively Art 1988); One Way EP (Lively Art 1989); Salad Days (Lively Art 1989).
DOMINION: Where Muses Dwell (KO City 1985}; The Oracle (KO City 1986); Melting the Emerald City (KO City 1988); Manhunt (KK 1990); Lost (KK1990); Battleground (Tragic Figures 1991).
PARADE OF SINNERS: Parade of Sinners (KO City 1986); This Is Where the Law Is (KO City 1988); Speech (KO City 1990).

CHRIS CONNELLY see Martin Atkins, Killing Joke, Pigface, Revolting Cocks

CONSOLIDATED

"We're just paranoic liberal white guys" - **Philip Steir**, 1990

Described in Spiral Scratch (amongst others) as a "white Marxist Public Enemy", Consolidated debuted with a self-titled EP in 1989. Since then, **Alan Sherburne**, **Mark Pistel** and Steir have become tagged "the unrelenting voice of an enraged generation", a **Jefferson Airplane** for the Industrial age. Encouraging their own live performances to become open debates with the audience, Consolidated is as much a political movement as a musical entity, a status they confirm (albeit less convincingly) on record.

Consolidated was formed from the wreckage of **Until December**, a Bay Area band who toured with **Gene Loves Jezebel**, released a long forgotten album on Columbia, but are best remembered today for their in concert cover of "Bela Lugosi's Dead".

Reconvening over a new social conscience (after "crack ethic careerism and total gang-rape ethics took over"), Consolidated dedicated their debut album to **Yussef Hawkins**, and having taken their message on the road, were rewarded with the success of 1991's Friendly Fascism.

"We weeded out some of the nihilistic aggression in the music. We don't think people should have to go out and get assaulted just as a matter of attending a cultural function", explains Sherburne. A CMJ Top 10-er, Friendly Fascism spawned several alternative hits, including "Unity of Oppression" (#4 on MTV's 120 Minutes' year's end best of chart), "This

Is Fascism" (#25 in Billboard's dance chart; #1 on Rockpool) and "Brutal Equation" (as above: Top 30 and Top 10 respectively).

Play More Music is the band's most fulfilled (and fulfilling) vision, although their unremitting pro-rights stance on issues as disparate as women, ethnic minorities, homosexuality and animals can become wearing in concentrated doses. A 12" single, "This Is Fascism", vibes up the earlier "Sexual Politics of Meat" or what could be construed as the ultimate self-parody.

Discography: The Myth of Rock (Nettwerk 1990). Friendly Fascism (Nettwerk 1991); Play More Music (Nettwerk 1992).

CONSTRICTION

"Constriction is long on frustration, short on tolerance and bound by circumstance." So says this California band on the jacket of its debut release, the thunderous In Vein

An anonymous two-piece, self-proclaimed cyber punks who proudly claim not to have driven legally in over a year ("fuck the police"), Constriction make heavy, and valid, use of sampling, but don't think "we really fit the stereotypical industrial sound."

Discography: In Vein (Constriction, 1992).

CONTAGION (formerly known as BIOHAZARD/PCB).

CONTROLLED BLEEDING

"If we're bruised and battered at the end of the show, then we've had fun" - **Paul Lemos**, 1992

From Boston out of New York City, Lemos and **Chris Moriarty**'s Controlled Bleeding underwent several false starts (the 1979 "Wall Of China" single on the indy Souptime label; a live at CBGBs album under the name of **Body Sink**) before Knees and Bones introduced them to an industrial audience in 1983. This time the cap fit, and Controlled Bleeding (Lemos the one constant in an otherwise shifting trio) went on to become one of the premier industrial bands, eclipsing even their earliest influences - **Cabaret Voltaire** and certain elements of **Joy Division** in terms of impact and inspiration. The arrival of **Joe Papa** introduced a more ethereal feel to the band's otherwise wired strivings.

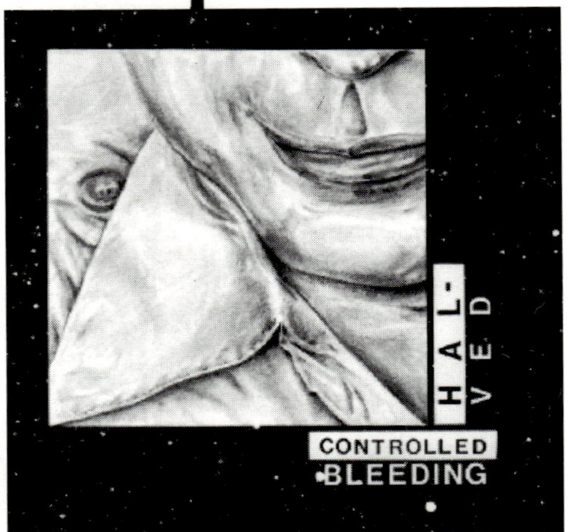

None of the band's first five albums (Body Samples followed Knees and Bones in 1984; the gentle Between Tides arrived in 1985, followed by Headcrack and Curd in 1986-87)) was released in the US. The band's homeland debut, then, was the bizarre Core, an album which found Controlled Bleeding experimenting within jazz and classical fields. To home audiences reared on the impassioned noise of the earlier records, this primarily instrumental excursion was something of a let-down - particularly as the band's European releases continued along the earlier, maniacal paths.

Songs from the Drain, released in Germany in 1988, and the Belgian Music from the Scourging Ground remain primal Bleeding. Some of the best of this continental material was included on the 1991 compilation Hog Floor; earlier Bleeding recordings resurfaced on Phlegm Bag.

This imbalance was partially corrected by the Songs from the Grinding Wall EP, a series of largely unstructured soundscapes, and the storming Trudge and Wound, albums which anticipated - and updated - much of **Ministry**'s early 90s' agenda. Wound, incidentally, appeared under the psuedonym **Skin Chamber** (originally **Fat Hacker**), hatched to allow the mothership "to move a little more creatively." Other side projects include **Pets for Meat**, described by Lemos as "sort of hip hop death music", and **Joined At The Head**.

Discography: Knees and Bones (Psychout 1983); Body Samples (Dossier 1984); Between Tides (Multimood 1985); Head Crack (Sterile 1986); Curd (Dossier 1988); Core (Subterranean 1987); Songs from the Drain (Dossier 1988); Music from the Scourging Ground (Play It Again Sam 1988); Songs from the Grinding Wall EP (Wax Trax 1989); Songs from the Ashes (C'est la Mort 1989); Les Nouvelles Mistiques de Chambre (Sub Rosa 1989); Trudge (Wax Trax 1990); Hog Floor (Subterranean 1991); Wound (**Skin Chamber** R/C 1991); Phlegm Bag Splattered (Dark Vinyl 1992); Penetration (Subterranean 1992); Golgotha EP (Soleil Moon 1992).

COP SHOOT COP

"Disco through a distortion box" - CSC on Ministry

Formed from the wreckage of mid-80s New York legends **Black Snakes**, the **Undead** and **Dig Dat Hole**, Cop Shoot Cop's amalgam of punk, sampling and noise-for-noise sake's psychosis is the missing link between pure-hearted Industrial music and the primitive hardcore agonies of early **Swans** and **Die Warzau.**

Cop Shoot Cop

Cop Shoot Cop

Cop Shoot Cop

Cop Shoot Cop - **Jim Coleman** (synth), **Tod Ashley** and **Jack Natz** (both bass), **Phil Puleo** (drums) - debuted in 1988 with the Japan-only Headkick Facsimile EP, a raw but exhilarating blast which set the stage for the band's debut album, Consumer Revolt. The band's third album Ask Questions Later is their first for Interscope.

Tod Ashley was a member of the shortlived **Shithaus**, with **Pussy Galore** percusslonist **Jon Spencer**.

Select discography: Facsimile EP (Supernatural Organization 1988); Piece Man EP (Circuit 1991); Consumer Revolt (Circuit, 1991); White Noise (1992); Ask Questions Later (Interscope 1993).

CORE see Chris & Cosey

COSEY FANNI TUTTI see Chris & Cosey

COUM TRANSMISSIONS see Throbbing Gristle

CREATIVE TECHNOLOGY INSTITUTE see Chris & Cosey

CRADLE OF THORNS

An exciting, young, fresh and insane band from the not-so-exciting town of Bakersfield, California. This industrial influenced "rage-rock" group is both enchanting and horrifying. Noted in the past for stage antics such as breaking jelly jars or setting free live crickets. Cradle of Thorns' sound features a powerful combination of male and female co-lead vocals. Psychotic and aggressive, the Thorns rage on.

Discography: Remember It Day (Castle Top 1990); Dysfunctional Brady Brunch/Behave (Cultivitation

CRASH WORSHIP

One may say it's Americas answer to **Neubauten**, San Diego's based Crash Worship put on a show that no one will forget, with people running around half naked with fire torches in their hand and a sound that will let your ears ring for a week. Formed in 1985 as a studio project inspired by the Japanese taiko troop **Kado**. In 1987 they started releasing material on Charnel House and the their own label called Alarma.

Discography: Flow (Alarma); Espontaneo (Charmel House 1991); The Science of Ecstasy (Alarma); Asesinos (Cold Spring 199[illegible]).

HELIOS CREED see Chrome

CRIMSON IVY

A Sunnyvale CA based trio, Crimson Ivy formed around 1987, and claim dedication "to the importance of recognizing the inter-relatedness of things", via a wall of electronic percussion and harsh, chanted vocals.

Discography: Spiritual Vision (CI 1988); Existence EP (CI 1990); We Are All Related EP (CI 1992).

DAVID CUNNINGHAM see Flying Lizards

CUPELO see Dome

CURRENT 93

Often pretentious, but generally thought-provoking, Britain's Current 93 combines a taste for Eastern musical forms with ideas reportedly culled from the works of **Aleister Crowley**. But throughout a decade-long career, Current 93 have hovered longingly on the fringes of the industrial scene, dipping tentative toes into the waters according to whom head Current **David Tibet** has most recently collaborated with. Thus the band's discography weaves itself through a variety of stylistic changes, with more recent incarnations lurching dangerously close to traditional folk themes.

Lashtal, released in Belgium in 1983, was a three track EP which also featured contributions from **John Balance** of **Coil** and **23 Skidoo**'s **Fritz Haaman**; two years later, Nightmare Culture turned over an entire album side to Coil, a sharp contrast to the eclectic chants and wails unleashed to hypnotic backing on Current 93's side. In between times, however, arguably the group's most important work appeared - Nature Unveiled, which combines Gregorian chants with Tibet's own interpretative meditations, and Dog's Blood Rising, an album in which Tibet's Eastern influences are allowed full reign, unimpeded by the use of non-original recordings. The album concludes with a **Simon and Garfunkel** medley, performed a cappella.

A combination of the backing tapes employed on these albums accompanies Tibet on 1986's Live at Bar Maldoror album; thereater, Tibet began his slow shift into more pastoral sounds, with In Menstrual Night and Imperium evincing an increased awareness of the folk genre. And while Dawn proved that he could still be vicious when he wanted to, Crooked Crosses for the Nodding God aka Swastikas for Noddy (a much-loved British children's character) showed that he didn't often want to. Earth Covers Earth and As The World Disappears, the final progression of Tibet's folkie leanings, are fascinating documents, and have been widely compared to 60s' esoterics the **Incredible String Band**. By comparison, Tibet's collaboration with **Nurse With Wound** is a welcomed jarring reversal of fortune. **(see Coil, Nurse With Wound, Boyd Rice)**

Discography: Lashtal (Laylah 1983); Nature Unveiled (Laylah 1984); Dog's Blood Rising (Laylah 1984); Nightmare Culture (with **Coil** Laylah 1985); In Menstrual Night (United Dairies 1986); Live at Bar Maldoror (Mi-Mart 1986); Dawn (Maldoror 1987); Imperium (Madoror 1987); Swastikas for Noddy (Laylah 1987); Earth Covers Earth (United Dairies 1988); Nylon Coverin' Body Smotherin' (with **Nurse With Wound** United Dairies 1988); Looney Runes (Dutro 1991); As The World Disappears (World Serpent 1992).

CHRIS CUTLER see Fred Frith

CYBERAKTIF

One time collaboration between **Skinny Puppy's Cevin Key** and **Front Line Assembly's Bill Leeb**.

Discography: Tenebre Vision (Wax Trax 1989).

HOLGER CZUKAY

Czukay embarked on a career in avant-garde music after witnessing a lecture by **Stockhausen** at Darmstadt in 1957. The pair composed several musical pieces together during the early 1960s, and in 1965 Czukay met fellow German **Irmin Schmidt**, a classically trained keyboard player who Czukay promptly baffled by inviting him to play a piece of "unplayable" music.

Holger Czukay

The two remained in contact and when, in 1968, Schmidt set about forming **Inner Space** (later to become **Can**), Czukay was an immediate recruit.

Czukay retained his independence, and in 1968 recorded his first album, Canaxis 5, with **Rolf Dammers**. A combination of pure electronics and ethnic music, the album was reissued in 1981.

However, his solo career did not truly get underway until 1979, following his departure from Can. Movies was a warm, but none-too-adventurous album, whose memory was easily erased by On The Way To The Peak of Normal, a devastating collection which included contributions from Can drummer **Jaki Liebe Zeit**, **PiL** bassist **Jah Wobble** and producer **Connie Plank**. (Czukay and Liebezeit were also involved in the shortlived **Phew**, whose eponymous album was released in 1981.)

Liebezeit and Wobble remained in tow for the adventurous <u>Wicked</u> (Virgin) in 1982, while 1983 brought a collaboration with Wobble and **U2** guitarist **The Edge** for the <u>Snakecharmer</u> mini album.

Working again with **Michael Karoli** and Lieberzeit, 1984's <u>Der Osten Ist Rot</u> and 1987's <u>Rome Remains Rome</u> bookended collaborations with the likes of **Cluster**, the **Eurythmics** and former **Japan** vocalist **David Sylvian** - their first collaboration was the excellent <u>Brilliant Trees</u> album in 1984, and the partnership reached its peak in 1988 with <u>Plight and Premonition</u>. (**see Can, Irmin Schmidt**).

Discography: <u>Canaxis 5</u> (Music Factory 1968); <u>Movies</u> (EMI 1980); <u>On The Way To The Peak of Normal</u> (EMI - 1982); <u>Der Osten Ist Rot</u> (Virgin 1984); <u>Rome Remains Rome</u> (Virgin 1987); <u>Radio Wave Surfer</u> (Virgin 1991).

DEATH METHOD

One of the new breed of electronic aggression to come out of L.A., along with **Spahn Ranch** their debut is highly rated for those into the harder side of techno, without being a Metal crossover.

Select discography: <u>Mentally Transmitted Disease</u> (1992 Q Music).

DEPECHE MODE

Emerging out of the London dormitory town of Basildon at the dawn of the synthipop boom, Depeche Mode cruised along as a comfortable, and generally mindless, pop group for three singles and one album. Then **Vince Clarke**, the group's principle songwriter quit (to form first **Yazoo** and then, as if there were any difference, **Erasure**).

Most people thought Depeche were finished, but they reckoned without **Martin Gore**. Whilst neither so glib, nor so commercially minded, as Clarke, Gore - under the tutelage of record company boss **Daniel Miller** - completely turned Depeche around, so that by the time of their third album, the excellent (and ingenuously titled) <u>Construction Time Again</u>, the group was being heralded as the most subversive pop group around. (Clarke was replaced, incidentally, by former **Hitmen** guitarist **Alan Wilder**, who joined vocalist **Dave Gahan** and **Andy Fletcher**.)

<u>Some Great Reward</u> cemented this reputation. Without once losing their grip on their teeny following (and a reputation to match), Depeche were now performing songs dealing with sexual perversions, religious falsehoods and sundry other adult themes.

Musically, too, Depeche had learned more fully the lessons taught by German and British industrialists past and present (**Einsturzende Neubauten** remain close to Gore's heart), than practically anybody else in the field, yet Depeche Mode continued enjoying massive singles success, at the same time as winning themselves a place in even the most unforgiving hearts. The singles "People Are People" and "Master and Servant", given dense, electronics-heavy remixes by **Adrian Sherwood**, increased the band's standing further.

<u>Black Celebration</u> and <u>Music For The Masses</u> showed the band continuing to advance down industrial pop lines, although the sameyness of some of the arrangements, and the increasing reliance on non-musical instruments

did begin to grate. The band themselves were aware of this, and took a lengthy sabatical, broken only by the release of a live album, 101. Anybody looking for an easy entry into Depeche Mode's career should take this one - the earlier People Are People (an America only compilation dating from 1984) and Catching Up With Depeche Mode (released in Europe as The Singles 1981-85) both give too much play to the earliest material.

The band returned to full power with 1990's Violator, an album which made up in inventiveness what it lacked in innovation. It was also a major hit, with one single from the set, "Personal Jesus", becoming the biggest selling CD5 in Warner Brothers' history.

Recorded after a lengthy sabbatical, 1993's Songs Of Faith And Devotion and the accompanying hit "I Feel You" are even more highly recommended.

Depeche Mode's continued mega-success has served to denigrate their contributions in the eyes of many people. But they remain industrial music's first, and greatest, crossover, and for that reason alone can never be easily written off.

Martin Gore and Alan Wilder (with **Recoil**) have both launched tentative solo careers. **(see Daniel Miller, Synthipop)**

Discography: Speak & Spell (Mute 1981); A Broken Frame (Mute 1982); Construction Time Again (Mute 1983); People Are People (Sire 1984); Some Great Reward (Mute 1984); The Singles 1981-85 (Mute 1985); Black Celebration (Mute 1986); Music for the Masses (Mute 1987); 101 (Mute 1989); Violator (Mute 1990); Songs Of Faith And Devotion (Mute 1993).

DEUTSCHE-AMERIKANISCHE FREUNDSCHAFT

Although they broke up as early as 1982, DAF joined **Einsturzende Neubauten** as prime movers on Britain's burgeoning industrial scene of the early-mid 1980s. Formed in Dusseldorf during the punk rush of 1977-78, DAF - **Robert Gorl** and **Gabi Delgado-Lopez** - began as just another gaggle of frenzied noise-merchants, shifting their focus towards calmer electronic dance music after hearing **Donna Summer**'s "I Feel Love".

Their approach, however, was never conventional. First heard in the UK on the 1979 compilation Earcom 3, the band then released Ein Produkt, a riot of synth- and tape-powered Heavy Metal; to an electronic audience raised on **Tangerine Dream** and **Kraftwerk**, it represented armageddon.

The band moved to London in 1980, signing to **Daniel Miller**'s Mute label. Die Kleinen und die Bosen was a considerably more refined work, and by the time of Alles Ist Gut, recorded with producer **Connie Plank**, the group was indeed moving in on **Kraftwerk**, albeit from a decidedly locker-room biology viewpoint.

A growing awareness of pure industrial music was demonstrated by the liberal use of found sounds - locomotives, etc. Gold Und Liebe and Fur Immer vary this approach considerably, with the latter embracing a number of otherwise alien musical forms wholeheartedly - rock'n'roll, funk, and so on. The best of these last three albums was subsequently compiled on the DAF retrospective.

Shortly after arriving in London, Gorl guested (as drummer) on **Eurythmics**' In The Garden debut. With DAF now (1983) on hold, and **Annie Lennox** a media star, their reunion on Gorl's Night Full of Tension album was widely anticipated, but generally disappointing. Delgado's Mistress, recorded with **Can** drummer **Jaki Liebezeit** and Connie Plank, was also poor. Sadly, a 1985 DAF reunion left behind no recorded work. **(see Donna Summer)**

Discography: Ein Produkt der Deutsche-Amerikanische Freundschat (Warning 1979); Die Kleinen und die Bosen (Mute 1980); Alles ist Gut (Virgin 1981); Gold und Liebe (Virgin 1981); Fur Immer

(Virgin 1982); DAF (Virgin 1988).

DEVO

"Are we not men?" asked five art school neurotics from Akron, Ohio in 1978, and when the answer came back, "We are Devo", the world was set (if not quite convinced) for an orgy of de-evolutionary preaching from possibly the daftest band to emerge from the entire post-punk firestorm.

Early singles on the Booji Boy label - "Jocko Homo"/"Mongoloid", "Be Stiff" and a **Residents-**like reworking of the Rolling Stones' "Satisfaction" (collected on the Be Stiff EP) - threatened to establish Devo as a genuine power; unfortunately, their first album Q: Are We Not Men offered nothing more exciting than the AOR staple "Whip It", and since then Devo's output has tended merely to irritate or, at best, titillate; their earliest recordings, however, captured first on the Mechanical Man bootleg and taken to extremes on "Satisfaction", remain essential pre-industrial strivings, despite latter-day attempts to nail them to other passing trends - Hardcore 1974-77 and so forth. Many of these would-be basement tapes have since been given full release by Rykodisc.

Ironically, the best Devo album since their early singles, was by another band entirely, **Clawhammer**'s Q: Are We Not Men? A: We Are Not Devo! (Sympathy for the Record Industry 1992), produced by Devo's own **Mark Mothersbaugh**, a note-for-note recapitulation of Devo's finest songs, played with more energy than the original Spudboys have managed in years. Mark Mothersbaugh has continued his career in T.V. theme and soundtrack work.

Selected discography: Mechanical Man (Nice 1978); Be Stiff EP (Stiff 1980).

DIATRIBE

A San Jose-based five piece which debuted on the Cyberflesh Conspiracy compilation, Diatribe (led by **Mark J**) scratch into view on the far side of **Ministry**, incorporating guitars and vocal treatments to far-ranging effect. Their debut EP "Nothing" was co-produced by **Lee Popa** and **Skinny Puppy** frontman **Ogre. (see Skinny Puppy)**

Discography: Nothing (Re-Constriction, 1992); Therapy (1992).

DIE KRUPPS

Die Krupps Formed in 1981 from the wreckage of German punk outfit **Male**, Die Krupps revolves primarily around multi-instrumentalist **Jurgen Engler** and on-off (currently on) collaborator **Ralf Dorp**, better known to English audiences for his work with mid-80s' surrealists **Propaganda. Lee Alters**, former guitarist for the **Heathens**, is also involved.

Increasing use of sampling saw Die Krupps progress from straightforward Teutonic thrash into a raging noise best experienced on 1992's Metal Machine Music (no relation!) EP, which also features guest appearances from The **Accused**. Also worthy of note are "Wahre Arbeit," a 1989 rerecording of a 1981 song, cut with **Nitzer Ebb**; and the five track A Tribute To Metallica, which **Metallica** themselves claim opened a new universe to their songs. "A song like 'Battery' was really thrashy," said Engels, "but with us he (**Metallica's Lars Ulrich**) could hear the melody for the first time!"

The best of Die Krupps is featured on the Past Forward compilation (see **Nitzer Ebb**).

Die Krupps

Selected discography: Stahlwerksinfonie (1981); Waher Lohn (1981); Volle Kraft Voraus (1982); Entering The Arena (1985); Machineries of Joy (1989); Metal Machine Music (Mute 1991); 91-82 Past Forward (Mute 1992); A Tribute to Metallica (Our Choice 1992).

DIE WARZAU

Chicago based Die Warzau, comprising Van Chrlstie and Jay Marcus, are very much a prototype of latter-day Ministry, relying heavily on driving percussion, melodic minimalism, creative sampling and overt political lyricism.

Die Warzau

The savage Disco Rigido catapulted Die Warzau to immediate attention, coincidentally allowlng the duo sufficient breathing space to fashion a deserving follow-up, Big Electric Metal Bass Face, an intense album of powerhouse industrial funk. Recommended.

Die Warzau

Discography: Disco Rigido (Fiction 1989); Big Electric Metal Bass Face (Fiction 1991).

DISCIPLINE

Sporting a mohawk, tattoos and pierced body parts, the image speaks for itself. Combining an Industrial/Rock fusion sound that booms and grinds ferociously, **Larry Law** puts on a pumping, grinding show for all new comers to the new Los Angeles Industrial rock scene.

Select discography: Discipline: Arbeit Macht Frei (1992 Discipline).

DOME

Bruce Gilbert and **Graham Lewis** remained together following the break-up of **Wire**, and throughout the early 1980s, maintaining a prodigious output of work all of which was characterized by its pioneering use of random electronics and white noise. Dome, Dome 2 and Dome 3, issued between 1980-81, are all recommended, as are the 1980 single "So"/ "Drop", and the Ends With The Sea 1980 EP, a collection of synthesized drones which, coupled with a brace of singles released under the name **Cupelo**, comprises the 8 Time compilation.

Another Gilbert-Lewis project, **P'o**'s 1983 album Whilst Climbing Thieves Vie For Attention also has its moments of inspiration.

MZUI/Waterloo Gallery remains the most adventurous of Dome projects in sonic terms, being in part a recording of a London art gallery's patrons, taped through hidden mikes.

A second MZUI album MZUI Australia, was recorded live via satellite from Sydney to London, with **Graham Bartlett** and **Russell Mills**.

Also of interest is the "Or So It Seems"/"Heart of Hearts" single, recorded under the ungrammatical alias of **Duet Emmo** with **Daniel Miller**, for the latter's Mute label.

A fourth Dome album, Will You Speak This Word was released in Norway only in 1983; Bruce Gilbert also released two solo albums, This Way and The Shivering Man, both of which are cut from more or less the same mold as early Dome. This Way To The Shivering Man, a compilation drawn from the first two titles, is the most immediately enjoyable introduction to Gilbert's often cruelly idiosyncratic musings. (A third Gilbert album, Insidings is of less intertest, being two lengthy dance pieces composed to order for choreographer **Ashley Page**.) **(see also Daniel Miller, Wire)**

Discography: Ends With The Sea (4AD 1980); Dome (Dome 1980); Dome 2 (Dome 1981); Dome 3 (Dome 1981); MZUI/Waterloo Gallery (Cherry Red 1982); Will You Speak This Word: Dome 4 (Uniton 1983); 8 Time (4AD 1988)
BRUCE GILBERT: This Way (Mute 1984); The Shivering Man (Mute 1986); This Way To The Shivering Man (Mute 1990); Insidıng (Mute 1991).

DUET EMMO see Dome

EBBINGHAUS

Haunting the fringe of the industrial dance scene, this Seattle duo - **Per Anderson** and **Kent Meyers** - debuted in 1992 with an excellent eponymous EP. Combining a latent '80s pop sensibility with a brooding density most noticeably inspired by an unholy collision between **Depeche Mode** and **Front 242**, Ebbinghaus' calling card is the demand, "We don't want anyone normal to find out about us."

Discography: Ebbinghaus EP (Iniquity 1992)

EINSTURZENDE NEUBAUTEN

If any band can be credited for welding together the entire would-be industrial scene of the early 1980s, it was Berlin conceptualists Einsturzende (whose name translates as Collapsing New Buildings and is not, therefore, so self-consciously amusing as, say, **Die Toten Hosen** - The Dead Trousers).

Whilst not completely espousing conventional musical instrumentation, the group's calling card is a dense wall of pounding, screeching and roaring, put together through the use of power tools, hammers and sundry other tools. Concerts generally turn into enactments of blacksmith's forges or demolition squads, and while vinyl can scarcely capture the full howling power of Neubauten, their catalog shows a dexterity which is a most surprising in its complexity and diversity.

The group debuted with the "Fuer den Untergang" single in 1979; together with tracks on a number of German compilations, this, like the double-pack Schwarz, showed Neubauten still to be coming to grips with their own unconventionality, offering up genuinely strong melodies alongside the burgeoning hammering.

Vocalist **Blixa Bargeld**'s collaboration with **Susie**, a 40 minute cassette called Valium, increased the band's own profile considerably, and by the time the band's debut album Kollaps appeared, it was impossible to judge Neubauten on musical merits, which had much to do with their adoption by the UK press around 1983.

Einsturzende Neubauten

Einsturzende Neubauten

Critically acclaimed, and signing to Some Bizzare, Neubauten became the band to watch throughout the country, yet only **Depeche Mode** truly latched on to their aural legacy, introducing to their arsenal many of the same sounds and devices as Neubauten was employing.

A brief union with **Lydia Lunch** and **Rowland S Howard**, on the run from the wreckage of **Birthday Party**, produced the magnificent Thirsty Animal EP; the following year, Drawings of Patient OT, the group's first UK album, was released.

Less harrowing than its predecessor, a consequence, perhaps, of Neubauten trying to move away from the slavering dogs who now hung onto their every clatter as though it were a burning bush, Drawings was a considerably more balanced album than Kollaps, and this is the face which Strategies Against Architecture, a 1986 compilation, turned to the world (a second compilation, Strategies Against Architecture II, was released in 1991). That said, there is nothing reassuring about a list of instruments which includes air conditioning ducts and a bridge!

The highlight of Strategies was the three live cuts included; 2x4 brought an entire Neubauten concert into your living room, a new ultimate in aural genocide. Halber Mensch, in 1985, was by comparison a relief, and to uneducated ears was certainly the most coherent Neubauten album yet. Which perhaps made it inevitable that the group would disintegrate almost immediately upon its release.

Blixa Bargeld

The breach was mended the following year, but Fuenf auf der Nach Oben Offenen Richterskala was a great disappointment, having more in common with "I Want More"-era **Can** than anything in Neubauten's own repertoire. The final Neubauten album, in 1989, rectified the failings of Fuenf somewhat, but Haus der Luege was still a far cry from what Neubauten fans expected.

The group reformed in 1992 for live work. **(see Depeche Mode, Synthipop)**

Discography: Schwarz (Zick Zack 1981); Kollaps (Zick Zack 1981); Thirsty Animal (with **Lunch/Howard**- GEMA 1982); Drawings Of Patient OT (Some Bizzare 1983); Strategies Against Architecture (Some Bizzare 1984); 2x4 (ROIR 1984); Halber Mensch (Some Bizzare 1985); Fuenf auf der Nach Obemn Offenen Richterskala (Some Bizzare 1987); Hans Der Luege (Some Bizzare 1989); Strategies Against Architecture II (Mute 1991); Tabula Rasa (Mute 1993).

ELECTRIC HELLFIRE CLUB

Thomas Thorn (formerly known as **Buck Ryder**) was the third member of **My Life With**

The Thrill Kill Kult on the Wax Trax label from 1988 to 1991. He left the group to form a new band when **Thrill Kill Kult** began pursuing a more mainstream disco sound. Prior to his involvement with T.K.K., Thorn worked largely with analog synthesizers on a number of projects, most notably the power electronics band **Slave State** (inspiration for the **Godflesh** album of the same name). **Thorn** currently handles the bulk of the programming duties along with acting as the group's frontman and vocalist. A voracious reader and longtime practitioner of the Black Arts, Thorn's interest in the occult and other esoteric subjects fuels the music's dark side, bleeding through the lyrical content into the sound itself.

Electric Hellfire Club

Originally a 3-piece, the group recently added a fourth member on electronic percussion. **Janna Flail** left sunny California in search of eternal damnation and found it in the Hellfire Club. A piercing enthusiast with a strong belief in enlightenment through pain, his fascination with primitive cultures has brought has brought a tribal elemant to the music's rhythms. His drumming is at times subtle but more often a brutal attack in which each song becomes a hypnotic ritual of unrestrained savagery. The addition of live percussion has given the music a fuller sound while Flail's untamed energy has given the band a new impetus. (see **My Life With The Thrill Kill Kult** and **Sleep Chamber**).

Discography: Burn Baby Burn (Cleopatra 1993).

BRIAN ENO

Eno has touched upon so many areas of modern/alternative rock, that to isolate his contributions to any one field is an impossible task. His pioneering role in **Roxy Music**, in which he fulfilled a role similar to, but far wider than, that enjoyed by **Hawkwind**'s **Dik Mik**, established him amongst the most colorful and innovative figures on Britain's turbulent

Eno

Glam Rock scene, and he took much of Roxy's original cartoon appeal with him when he went solo, in 1973.

For the best of his work with Roxy, "If There Is Something" from the bootlegged John Peel session of 1971, "The Bob (Medley)" from the band's 1972 self-titled debut album, and such b-sides as "The Numberer" and "The Pride & The Pain" are most heavily reflective. "Sultanesque", a droning post-Eno cut, indicates that his influence continued to hang heavy over group leader **Bryan Ferry**; on the b-side of 1974's "Love Is The Drug" single, it vies with **Kraftwerk**'s "Autobahn" as the first industrial song ever to make it onto British jukeboxes!

"Seven Deadly Finns" and "The Lion Sleeps Tonight", singles otherise available only on the Rarities disc included within the eleven disc Working Backwards 1983-73 boxed set, are as idiosyncratic as anything else Eno has recorded, although his first two albums - Here Come The Warm Jets and Taking Tiger Mountain By Strategy - remain essential listening, anticipating as they do so many of rock's future obsessions. Punk, Goth and Industrial musical forms can all trace their lineage through these seminal albums.

With Another Green World, Eno's disillusion with "conventional" rock formats first started to become evident; Before and After Science was the last of his so-called "pop" albums, but even here he was moving in new directions. More Blank Than Frank is Eno's own selection of tracks from these first four sets.

But it is as a producer that Eno has had the greatest impact. His work with **Robert Calvert** produced the startling Lucky Lief & The Longships set; he was associated with the early **Television**, **John Cale**, **John Foxx/Ultravox**'s debut album, and has even lifted **U2** and **David Byrne** out of their own musical troughs.

The icy core of **David Bowie**'s Low and Heroes, and **Iggy Pop**'s Bowie-fired The Idiot, all released in 1977, mark the culmination of Eno's earliest flirtations with the industrial sound. No examination of Eno's impact on industrial music is complete without these albums.

Eno's ambient albums, the first of which, Discreet Music, appeared in 1975, and which initially borrowed as much from the likes of **Can** and **Cluster** as they gave to later generations, have their own specific following, which frequently moves beyond that afforded his earlier records; of greater interest are the two sets he recorded with members of Cluster - Cluster and Eno in 1977, and After The Heat in 1978. Old Land compiles tracks from both.

More important, however, are No Pussyfooting and Evening Star, two albums of pure electronics recorded with **King Crimson**'s **Robert Fripp**. Challenging and uncompromising, the two records were widely ignored upon release, but they remain massively influential. Notable is the inclusion of excerpts from No Pussyfooting on the 1992 King Crimson live set, The Great Deceiver. **(see David Bowie, John Cale, Robert Calvert, John Foxx, Robert Fripp)**

Selected discography: Here Come The Warm Jets (Island 1973); Taking Tiger Mountain By Strategy (Island 1974); Another Green World (Island 1975); Before and After Science (Island 1977); More Blank Than Frank/Desert Island Selection (EG 1986).
With ROBERT FRIPP No Pussyfooting (Help 1974); Evening Star (Help 1975).
With CLUSTER: Cluster and Eno (Sky 1977); After The Heat (Sky 1978); Old Land (Relativity 1985).

ETHYL MEATPLOW

If your imagination were a household appliance, what would it be? The all-purpose answer, the original power tool, where nuts and bolts meet flesh and fantasy–Ethyl Meatplow.

Androgyny rears its spinning head while a disturbing funk rhythm cuts a swath through the dance floor. The combined energies fo **Carla Bozulich, Biff Barefoot Sanders** and **John Napier**, have been unleashed on a wide range of stunned audiences both cult in Los Angeles and the curious throughout the nation. Everthing about Ethyl Meatplow is live-like tossing a hot plugged-in sampler into the bathwater. Power trio? Maybe...but not your regular scenario. You get three people on high with no one ever taking the back burner. Powered by computers, machines, samplers and various tinkerings. Songs drenched in delusion, conspiracy theories, frustration and newfound freedoms is the order of the day here.

Ethyl Meatplow

Discography: Spasm Sampler (Spasm); Dancing With Porkface- 7" E.P. (Spasm); MOTIV-12" (Spasm) Whore/Queenie-7" (Sympathy For The Record Industry); Happy Days Sweetheart-CD (Dali).

EXCESSIVE FORCE see KMFDM, My Life With The Thrill Kill Kult

EXECUTIVE SLACKS

A Philadelphia-based trio, Executive Slacks was the ferocious fusion of industrial hardcore and **Residents-**style Glam Rock - their rendering of Gary Glitter's seminal "Rock and Roll", on their final Fire and Ice album was the mode from which British acid-industrialists **KLF** constructed their own chart-topping "Doctoring the Tardis" remake, but would not have been out of place in **Throbbing Gristle**'s repertoire.

The band's debut, the Executive Slacks EP, set the stage for subsequent releases, and is included, alongside a second EP, in its bone-crunching entirety on You can't Hum When You're Dead, a blistering statement of intent which borrowed much from producer **Youth**'s old mates in **Killing Joke**. This unrelenting hammering was kept up across the aptly titled Nausea, by which time the joke was beginning to pale, hence the transitionary introduction of sundry other musical forms - most unsuccessfully Hip Hop - into the brew.

Fire and Ice, however, lifted only what worked from its predecessor, and proved a fitting swansong for a most entertaining band. Executive Slacks split in 1986. (see **Tubalcain**).

Discography: You Can't Hum When You're Dead (Fundamental 1984); Nausea (Fundamental 1985); Fire and Ice (Fundamental 1986).

FAD GADGET see Frank Tovey

THE FALL see Punk Rock

FALL OF BECAUSE see Godflesh

FAUST

Although the German band had been recording since 1971, Faust became generally known the following year when they accompanied **Slapp Happy** on that group's debut album, Sort Of. The following year, Faust's UK label Virgin released the budget priced Faust Tapes, a compilation of out-takes from the band's first two records, Faust and So Far. (A similar accumulation of floor sweepings was employed for the very posthumous Munich and Elsewhere and The Last LP collections.)

Pioneering the long-term use of musical effects hitherto restricted to isolated moments in mainstream rock - heavy distortion, loud droning fuzz tones and instruments which wander in and out of the mix seemingly at random, Faust were initially described as a Teutonic **Mothers of Invention**, a tag which So Far, with its bewildering mood and method swings, gleefully lived up to.

The more atmospheric Tapes, however, sent the band and its new-found audience moving back into the realms of almost pure electronics, but the benefits were double-sided. With **Kraftwerk**, **Can** and **Tangerine Dream** now also ranking high in people's affections, Faust found themselves competing not only with their own reputation, but also their countrymens'. "Kraut Rock", the best-remembered track from the disappointing Faust IV attempted to put the record straight by launching into a lengthy musical parody of those other bands. It failed, and Faust broke up shortly after. (**see Kraut Rock, Slapp Happy**)

Discography: Faust (Polydor 1971); So Far (Polydor 1972); The Faust Tapes (Virgin 1973); Faust IV (Virgin 173); Munich and Elsewhere (Recommended 1986); The Last LP (Recommended 1988).

FEAR FACTORY

The self-described ultimate union of industrial, death metal and grindcore, Fear Factory formed in Los Angeles around **Dino Cazares** and vocalist **Burton C. Bell**, drummer **Raymond Herrera** and bassist **Andrew Shrives** in October of 1990. New remixes of Fear Factor have been taped by industrial dance masters Front Line Assembly.

Select discography: Soul of a New Machine (1992 Roadrunner).

FLESH VOLCANO see Foetus, Soft Cell

FLYING LIZARDS

As the last embers of punk rock fell on a disinterested hearth, **David Cunningham** looked around at the success of LA's **The Dickies**; listened hard to the work of **The Residents** and **Daniel Miller**'s **Silicon Teens** enterprise; and figured, "Hey, I can do that." Teaming up with the mysterious **Deborah Lizard**, the result was the Lizards, the first - and hopefully the last - industrial novelty band.

Their debut, a hammer-heavy retread of "Summertime Blues", and the similary quirky "Money" follow-up, set the stage for both half of the first, and all of the final Lizards album; The Flying Lizards, however, was further distinguished by Cunningham's attempts to out-dream **Tangerine Dream**, and the whole thing is best forgotten.

A grandiose Cunningham solo album, Grey Scale, fell similarly short of its mark, and is best described as **Keith Emerson**'s lost "Variations on a theme of **Terry Riley**'s brother", or something equally unpalatable.

After such dire beginnings, then, the Lizards might have been excused from the room. Instead, Cunningham turned all expectations around with The Fourth Wall, a devastating collection which, in later years, would appear to have anticipated, and outdone, the Anglo acid scene of a decade later. Certainly bands like **the Orb** would do a lot better if they knew that what they are doing had already been done! Guest musicians include a bemused sounding **Robert Fripp** and **Patti Palladin**.

Top Ten, too, proved a considerably more enjoyable album than may have been expected; revamping 50s' and 60s rock classics to a pulsing electronic beat, Cunningham succeeds in raising his work far above the simple novety-nuisance values which plagued The Flying Lizards, and creating an album of lasting worth.

Discography: The Flying Lizards (Virgin 1979); Grey Scale (David Cunningham: Piano 1980); Fourth Wall (Virgin 1981); Top Ten (Statik 1984).

FOETUS INC.

"I try to view my work as an entity unto itself, not that I am like one of many, or that I'm part of this big fucking dinosaural bureaucracy who are trying to sell jeans, or sell themselves or whatever. If I ever considered myself part of it I think I'd kill myself" - **Foetus**, 1989

It sounds like a business empire, a steel and glass monstrosity towering above the town, its corporate identity flashing on and off all night. Foetus Incorporated... but who'd dare to guess what they're doing in there?

Stone-age metal retreads of **Blue Oyster Cult**'s "Don't Fear The Reaper"...grinding horror reruns of "Why Don't We Do It In The Road"...and "ethnic music from a civilization yet to be invented." Midway through 1991, **Jim Thirlwell** a.k.a. **Clint Ruin** a.k.a. Foetus had no less than three new releases in the stores: Petal To The Metal, under the name of **Wiseblood**; Quilombo, credited to **Steroid Maximus**; and the Don't Fear The Reaper EP, pairing him with **Lydia Lunch**.

Foetus

Another two albums, Gonwanaland and Male - were forthcoming, a body of work which by itself would have been sufficient to establish

Foetus amongst the most prolific, not to mention creative, figures around. The fact that they merely added to an already immense catalog simply boggles the imagination further.

Australia-born Thirlwell arrived in England in 1978, falling from the same boatload of Australian punk invaders as **Birthday Party** and **The Saints**. But he swiftly set himself apart from the rest of the scene - as his choice of aliases revealed: **Foetus Under Glass**; **Foetus Art Terrorism**; **Scraping Foetus Off The Wheel**; **Foetus Over Frisco**; **Philip & His Foetus Vibrations** and - is this the most tasteless name in rock? - **You've Got Foetus On Your Breath**. What sort of person is this?

The names shocked, Thirlwell admitted. They turned people off before they'd even heard the music. They were the price of admission. If you could survive the name you were ready for the music, a frenzied clash of performance art and practical dance music, which was as compulsive as it was disturbing. "Wash It All Off", the 1981 single which introduced the Breath alias to the public, remains Foetus' most startling statement of intent; pounding percussion, tortured rhythms and over it all Foetus snarled and yowled through the one musical soundtrack the Eighties truly deserved.

In 1982 Thirlwell linked with **Marc Almond** for a one-off performance of **Suicide**'s "Ghostrider" on British television. It was a weird pairing: Almond was one half of **Soft Cell**, darlings of Britain's disco Top 20; Thirlwell a cult known only for his bizarre taste in aliases.

But it worked, and before the year was out the pair had brought their act to America, linking with **Birthday Party**'s **Nick Cave** and **Lydia Lunch** to perform as the **Immaculate Consumptives**. (Thirlwell and Almond regrouped for the "Slut" single in 1988, this time in the guise of the **Flesh Volcano**.) Other Foetus collaborations include work with acts as disparate as **Nurse With Wound** and Brit sugar-rockers **EMF**.

Foetus, meanwhile, rolled on. Signed to Soft Cell's Some Bizzare label, he flirted briefly with the UK pop mainstream - the Almond association had opened his name, if not his vision, to mass-consumption. But 1983's Hole and '85's Nail, were juggernauts careening through residential streets; the kids couldn't dance to "The Theme From Pigdome Come". Foetus moved to New York City.

1987 brought the somewhat chilling notion of the **Foetus All Nude Review**, a (thankfully unfulfilled) promise which delivered a stunning mini-album, Bedrock. The following year, **Foetus Interruptus** released the shattering Thaw, a set so savage it could never be recaptured on stage. Thirlwell acknowledged that much when he went out on tour. He was billed as **Foetus Corruptus**.

The highlight of the show was "Faith Healer", **Alex Harvey**'s classic concert-opener (Thirlwell cites Harvey as an influence and an idol). It was an inspired choice. For there are precious few people around who could more easily fit into that role. Spray gunning his own personal philosophies, daubing them like graffiti then leaving them hanging unanswered in the air, Foetus is one of rock's last remaining visionaries, its last surviving exorcist. Like the song says, "Your spirit never has to grieve - all you have to do's believe." One show, at CBGBs was recorded for the double live Male, credited to **Foetus in Excelsis Corruptus Deluxe**.

Sink is a double album's worth of Foetus singles, out-takes, rarities and remixes which debuted Foetus Inc. in 1990. **(see Coil, Nurse With Wound, Soft Cell)**

Selected discography: Deaf (You've Got Foetus Self Immolation 1981); Ache (You've Got Foetus Self Immolation 1982); Hole (Scraping Foetus Self Immolation 1984); Nail (Scraping Foetus Self Immolation 1985); Dirtdish (Wiseblood K.422 1986); Bedrock EP (All Nude Foetus Self Immolation 1987); Thaw (Interruptus Self Immolation 1988); Stinkfist EP (**Ruin/Lunch** Widowspeak 1988); Rife (Interruptus Self Immolation 1989); Sink (Self Immolation 1990); Petal To The Metal (Wiseblood K.422 1991); Quilombo (Steroid Maximus Wax Trax 1991); Don't Fear The Reaper EP (Ruin/Lunch Big Cat 1991); Gonwanaland (Big Cat 1991); Male (Foetus Corruptus Big Cat 1992).

JOHN FOXX

They lost it in the end, but for a moment there, **Ultravox** seemed set to shape into something very precious. Their eponymous debut album borrowed liberally from **Doctors of Madness**, **David Bowie** and producer **Eno** to create a shimmering landscape of pre-punk neuroses, its impact deadened only by the record company's failure to market it as anything but post-punk outrage. Songs like "Saturday Night in the City of the Dead" (the Doctors' "Mainlines" revisited) and the debut single "Young Savage" cast Ultravox far beyond even the wildest punk slung stone, and this was the direction the band continued to move over two subsequent albums.

John Foxx

Of these, HaHaHa includes the remarkable "Rockwrok", a fleeting moment of robotic light relief most effectively heard (using the CD programmer) in between the album's two key tracks, "Hiroshima Mon Amour" and the maniacal "Fear In The Western World"/"Distant Smile" medley.

Systems of Romance, the third Ultravox album, is weaker, but producer **Connie Plank** does pave the way for both his and Foxx's later work. Thus, while Ultravox itself eventually became a vehicle for the wan pontifications of former teenybop idol **Midge Ure**, Foxx - the original band's vocalist and lyricist - admitted that his vision had been thrown seriously off-kilter and, lurching into the now-burgeoning New Romantics movement, launched upon a solo career which simultaneously conveyed asexual alienation and a warm affection for machines as living beings.

Metamatic, his first solo album, is the most heavily influenced by **Gary Numan**'s now sagging pretensions, and for a time, Foxx seemed set to follow Numan all the way to the top of the British chart.

But the songs-most-likely-to, "Underpassing" and the No-one Driving EP stopped short of the magic 30, and Foxx retired to compose an entire album around the one track on Ultravox which didn't hold out hope for the future, "My Sex" (strangely, a favorite amongst his predominantly college-age following).

The Garden, though, dispensed with its predecessor's stark acoustics to present an album of almost **Yes-**esque lushness, over which Foxx's melancholic vocals (and themes!) sounded ill at ease. The Golden Selection and In Mysterious Ways moved even further from the harsh, doomladen mechanics of his earliest, and by a long chalk, best work. **(see Brian Eno, Synthipop)**

Discography: ULTRAVOX Ultravox (Island 1976); HaHaHa (Island 1977); Systems Of Romance (Island 1978); The Peel Sessions (Strange Fruit 1987).
JOHN FOXX Metamatic (Metal Beat 1980); No-one Driving EP (Virgin 1980); The Garden (Metal Beat 1981); The Golden Section (Metamatic 1983); In Mysterious Ways (Metal Beat 1985).

Robert Fripp

ROBERT FRIPP

The self-styled enigma Fripp first came to note with (Mike) **Giles**, (Pete) **Giles & Fripp**, a pleasant but distinctly unexciting folkie trio who cut one loser album in 1968. Augmented by **Ian MacDonald** and **Pete Sinfield**, and with **Greg Lake** replacing Pete Giles, **King Crimson** debuted in London in April 1969, recipients of an immense, and immensely succesful hype. They followed this with an appearance at the **Rolling Stones**' free concert in Hyde Park that summer, and in the fall, their debut album Court of the Crimson King was released to unanimous acclaim.

In many ways, Crimson's development paralleled that of **Van Der Graaf Generator**, the only other UK band of both commercial and critical standing who flirted constantly with themes from beyond the rock mainstream. In The Wake of Poseidon, Islands and Lizard all feature their fair share of adventure, although it is perhaps the live Earthbound which best highlights Crimson.

Poorly recorded on their 1972 European tour, and featuring lengthy, sub-metallic versions of the anthemic "21st Century Schizoid Man" and "Groon", Earthbound captures the otherwise unrecorded line-up of Fripp, Sinfield, **Boz Burrell** (bass), **Ian Wallace** (drums), and **Mel Collins** (reeds) taking the band's sound to extremes which would not be recaptured until **Cabaret Voltaire** began first venturing onto the road. (A second live album, USA, sadly failed to meet its predecessors' nightmarish standards, despite being recorded on the back of three excellent studio sets, Larks Tongue in Aspic, Starless & Bible Black and Red. The early-mid 70s band was also documented on the four CD live set The Great Deceiver, which also features excerpts from No Pussyfooting as intro and outro themes, plus two improvisational pieces, "Some Pussyfooting" and "Some More Pussyfooting".)

Although the King Crimson story has continued intermittently to this day (a reformed line-up is expected to start work during 1993), Fripp's outside commitments, most notably alongside **Brian Eno** (the No Pussyfooting/Evening Star experiments), **Flying Lizards** (their final album), **Peter Gabriel** (the former **Genesis** frontman's eponymous second album) and the **League of Gentlemen** (with ex-**XTC** keyboard player **Barry Andrews**) has long since taken precedence, to the extent that Crimson's undeniable impact on both the pre- and post-punk music scenes has now been forgotten, a happenstance which might be unjust but is not unjustified.

Beginning with 1979's Exposure album, Fripp has ridden his earlier vision wherever it might take him, with his tape-loop guitar "Frippertronics", a trademark which no guitarist of the last decade has come close to emulating. Guest appearances with **David Bowie**, **Blondie** and, of course, wife **Toyah** have sent Fripp's reputation and sound spiralling into quite unexpected avenues, so that today there are few areas of popular (and unpopular!) music which can claim to be wholly untouched by the man's influence. **(see David Bowie, Brian Eno, Flying Lizards, League of Gentlemen, XTC)**

Selected discography: KING CRIMSON Court of the Crimson King (Island 1969); In The Wake of Poseidon (Island 1970); Islands (Island 1970); Lizard (Island 1971); Earthbound (Help 1972); Larks Tongue in Aspic (Island 1973); Starless & Bible Black (Island 1974); Red (Island 1975); A Young Person's Guide to King Crimson (Island 1975); The Great Deceiver (EG/Caroline 1992).
ROBERT FRIPP: No Pussyfooting (with Brian Eno Help 1973); Evening Star (Help 1975); Exposure (EG 1979); Under Heavy Manners/God Save the Queen (EG 1980); Let The Power Fall (EG 1981); The League of Gentlemen (EG 1981); God Save The King (with League of Gentlemen EG 1985); Live (with League of Crafty Guitarists EG 1986).

FRED FRITH

Alongside **Slapp Happy** and, to a lesser degree, **Hatfield and the North**, guitarist Frith's band **Henry Cow** spearheaded the British prog rock scene's descent into the realms of the avant-garde, a band who combined American jazz improvisations with European art music to create a series of albums which remain truly unique.

That only two albums - Legend and Unrest truly testify to the original band's brilliance is unfortunate, but has to a degree been rectified by subsequent recordings by founder member Frith. Also of note are the two albums which paired Cow with Slapp Happy, after which the latter's vocalist, **Dagmar Krause**, joined Henry Cow in time for the live Concerts album. Frith also contributed half a dozen dischordant viola arrangements to Velvet Donkey, a collection of idiosyncratic anecdotes, vignettes and silly songs by **Ivor Cutler**.

Henry Cow broke up in 1976, reforming two years later (without Krause) to record Western Culture. In the meantime, Frith had released his first solo album, Guitar Solos; and put together the **Art Bears** with drummer **Chris Cutler** and the errant Krause. The three albums they recorded between 1978 and 1981 stand as unimpeachable monuments to the band's far-sightedness; although Krause's high-pitched speed-singing is often disorienting, Cutler's booming percussion and Frith's array of treated stringed instruments is constantly challenging. From the Winter Songs sessions, the "Rats & Monkeys" single is excellent.

The joint Frith/Cutler Live in Prague and Washington album remains a highpoint in either musician's career, and had Frith not then decided to relocate to New York, there is no telling what may have happened next. In any event, Cutler and Krause continued to work together within the introspective **News From Babel**, with Krause continuing Babel's flirtation with **Brecht**ian cabaret across her two solo albums.

The second and third volumes of Guitar Solos, upon which Frith was but one of several players, appeared in 1979, shortly before his first collaboration with that other avant-garde guitar great, **Henry Kaiser**. The pair recorded two extreme albums together, then took the show onto the road to produce the in-concert With Enemies Like These, Who Needs Friends?, a startling showcase which captures both men at their best. (More of Kaiser's idiosyncracies can be found on the 1992 compilation Lemon Fish Tweezer).

Midway through the Art Bears' lifespan, Frith recorded Gravity for **The Residents**' Ralph label, a strong album highlighted by a weird instrumental version of the old Motown chestbeater "Dancing in the Streets". The following year, Speechless emphasised Frith's own powerful interest in classical music, but remained within a strong rock framework, similar to that mined by Slapp Happy. It remained Frith's solo highwater mark - at least in the studio - even after the release of Cheap at Half the Price, an entertaining but patchy album which Frith recorded in his home four track studio.

A rudimentary approach to recording was also evident in Live in Japan, which is nothing short of two discs full of raw noise gouged out on a succession of (often homemade) guitars. Taking **Jimmy Page**'s violin bow technique to sometimes absurd limits, Frith attacks his strings with a variety of common - and not so common - objects, creating an album which remains the most aurally abrasive in his catalog. Further live material, incidentally, can be found on the often difficult Step Across The Border compilation. Another collection, Guitar Solos Complete, features all of Frith's first solo album, plus his contributions to Guitar Solos 2 and 3.

His two albums with the blistered folk-oriented **Skeleton Crew** aside, the decade since Live in Japan has seen Frith step back from the edge somewhat, to concentrate upon film and dance scores. Often employing some unusual musical techniques and textures, Frith's last two albums are nevertheless no more than sketches, crying out for greater definition. **(see Residents, Slapp Happy)**

Selected discography: FRED FRITH Guitar Solos (Caroline 1974); Hopes and Fears (with Art Bears Random Radar 1978); With Friends Like These (with **Henry Kaiser** Metalanguage 1979); Winter Songs (with Art Bears Ralph 1979); Gravity (Ralph 1980); Speechless (Ralph 1981); The World As It Is Today (with Art Bears Re 1981); Live In Japan (Recommended 1982); Who Needs Enemies? (with **Henry Kaiser** Metalanguage 1983); Cheap At Half The Price (Ralph 1983); Live in Prague and Washington (with Chris Cutler Re 1983); Learn To Talk (with Skeleton Crew Rift 1984); The Country of Blinds (with Skeleton Crew Rift 1986); With Enemies Like These, Who Needs Friends? (with **Henry Kaiser** SST 1987); The Technology of Tears (SST 1988); The Top Of His Head (Crammed 1990); Step Across The Border (East Side Digital 1990); Guitar Solos Complete (East Side Digital 1991).
HENRY COW Legend (Virgin 1973); Unrest (Virgin 1974); Desperate Straights (with Slapp Happy Virgin 1975); In Praise of Learning (with Slapp Happy Virgin 1975); Concerts (Caroline 1976); Western Culture (Broadcast 1979).

FRONT 242

"We want to reach everyone who is open to our music" - **Richard 23**, 1991

Despite having long provided a ready market for industrial and techno experiments which British and US outlets were unwilling to touch (the pioneering Laylah label for instance), Belgium's own contributions to the genre were restricted to the likes of **Telex**, a synthipop duo whose achievements include working with **Sparks** and entering the dreaded Eurovision Song Contest.

Front 242

So few people took Front 242 (or **Prothese** as they were originally known) at all seriously when they first emerged in 1981, a French-speaking **Ministry**, whose needle was well and truly struck in the No Sympathy groove. A succession of dancey 12" singles, beginning with "Principles"/"Body to Body" (New Dance) and filler-laden albums in their native land passed unnoticed, but like **Depeche Mode**, the worm was soon to turn. They even stopped playing 15 minute gigs!

The missing link in the pop vision of **Jean-Luc de Meyer**, **Patrick Codenys** and **Daniel Bressanutti** appears to have been **Richard 23**, who joined the group in 1983, after their debut album Geography.

But even he was unable to do more than sharpen the group's edges a little; it was four years before the modern Front 242 finally emerged, a bristling monster whose first "born again" album was Official Version, a record which the band claimed was the first to capture the true nature of their stage show.

Signing to Wax Trax, who promptly reissued the band's Geography debut, the Endless Riddance and Politics of Pressure sets, the limited edition Live in Chicago, an almost-instrumental EP No Comment, and a compilation drawn from them all, Back Catalog, Front 242 immediately began picking up press in the US for their vicious brand of industrial dance; they repaid their new-found admirers with Front By Front, a storming tour de force which erupts out of the record player with the finesse of an anti-personnel bomb.

The somewhat disappointing "Never Stop" single notwithstanding, Front 242 maintained the barrage onto Tyranny For You, by which time their reputation had seen them transfer allegiance to the multi-national Sony Corporation, proof that not all Europeans have to sound like Golden Earring and Nena to make it big these days. 1993 sees the release of two new albums on Epic as well as partaking in the Lollapalooza tour. (**see Revolting Cocks**)

Selected discography: Geography (Wax Trax 1983); No Comment (Wax Trax 1985); Official Version (Wax Trax 1987); Front By Front (Wax Trax 1988); Back Catalogue (Wax Trax 1985) Tyranny For You (Epic 1991); Up Evil (Epic 1993).

FRONTLINE ASSEMBLY

"Can you imagine what things will be like 100 years from now? All the futuristic stuff we're doing now will be obsolete" - **Bill Leeb**, 1992

Formed by ex-**Skinny Puppy** man **Bill Leeb** (he quit in 1986 in a dispute over "creative freedom"), plus fellow Canadian synth player **Rhys Fulber**, Frontline Assembly was founded as an experiment in electronic dance music, fused with samples and sound bites.

The 1988 Gashed Senses album is generally spectacular; its highlight, "No Limit", also appeared as a seven minute 12" single, presented in "Damaged Goods Remix" form, acknowledging the band's debt to **Gang of Four**.

Front Line Assembly

Early albums - State of Mind and Initial Command Fruehstick - struggled to make an impact; by the time of Caustic Grip, however, Frontline Assembly was definitely living up to its creators' own dreamplan: the more song-oriented Tactical Neural Implant was even stronger, certainly defying **Nine Inch Nails** frontman **Trent Reznor**'s complaint that the band was "monotonous, boring, uninspired bullshit." Leeb was unperturbed. "It wouldn't make me feel any better to see him lying on the floor in a pool of blood," he replied. **Bill Leeb** and **Rhys Fulber** also record as **Noise Unit**, **Intermix** and **Delirium.** (**see Skinny Puppy**).

Selected discography: Gashed Senses and Crossfire (Wax Trax 1988); Caustic Grip (Wax Trax 1989); Tactical Neural Implant (Roadrunner 1992); Euphoric EP (as Delirium 1992); Convergence (Third Mind, 1992). NOISE UNIT: Frequency Response (Antler 1990); Strategy Of Violence (Dossier 1992).

PETER GABRIEL see Robert Fripp

GEKO

The two women who form this highly acclaimed Los Angeles band probably hold the distinction of being the world's only all female Industrial Goth group. At first listen this duo might remind you of early **Christian Death**. Gloomy, sensual, and sonic describes the music the music of Geko which features **Sarah Folkman** (Vocals, bass), **Carrie McNinch** (guitar) and their faithful drum machine.

Discography: Probing The Gash In Her Head (Open Records 1991); Join My Pretty World (Open Records 1992).

GODFLESH

"[Our music] is about paranoia, alienation, a feeling of being trapped.... It's the sound of extreme shittiness, basically" - **Justin Broadrick** 1990

Emerging from stints within primal thrash monsters **Napalm Death** and **Head of David**, Birmingham, England based guitarist **Justin Broadrick** spent several months playing his **Swans** albums at the wrong (slow) speed, at the same time as working alongside drummer **Paul Neville** and bassist **G Christian Green**, and some tape loops and a drum machine, in **Fall of Because**.

Following his departure from Head of David, Broadwick convinced the other two to change their name to **Godflesh**, a band which has been likened to everything from Pornography-era **Cure** on quaaludes to a very protracted course of root canal work. Broadrick's vocals, incidentally, usually approach the listener from the other end of a very long, dark tunnel.

Godflesh, the band's debut album, and the harsher Streetcleaner, remain their finest moments; also recommended (and due for compilation soon) is the collection of singles and EPs which have occupied the band's time since they discovered dance music late in 1990. Included amongst these was a stunning version of **Loop**'s "Straight To Your Heart" (Clawfish).

Neville quit following a UK tour with Loop; he was replaced by Loop's own **Robert Hampson** (Napalm Death drummer **Mick Harris** also played a handul of shows with the band), and in this form recorded Pure, amongst whose tracks nestles a 21 minute feedback drone, through which floats an unrecognizable **Beach Boys** sample. Despite the departure of Hampson following a 1992 European tour, the band spent the summer touring with **Skinny Puppy**. (Hampson is currently putting together a new project, **Main**, with help from Green.)

Sex Gang Children

Selected discography: Godflesh (Silverfish 1988); Streetcleaner (Earache 1990); Slavestate EP (Earache 1991); Pure (Earache 1992); Cold World (Relativity 1992).

GOTH

Although it unquestionably had its roots in **Siouxsie and the Banshees**' darker meanderings (much of Switch, the "Voices" b-side, etc), and was born out of **Bauhaus**' seminal "Bela Lugosi's Dead", the Goth scene which took the early 1980s by the throat (and hasn't relaxed its skeletal grip since) is a rare example of an entire musical genre coming to (half-) life with neither precedent nor prompting to propel it along.

Bauhaus frontman **Pete Murphy** still insists "Lugosi" was recorded as a joke; but who got the last laugh? The band, for pulling the wool over so many people's eyes? Or the people, for making Bauhaus, Murphy and the spin-off **Love and Rockets** live with the punchline for so long?

Centered around the ubiquitous Batcave, a London club whose decor was lifted directly from a 1960s Hammer Horror flick, words such as emergent, burgeoning and embryonic simply cannot be applied to the Goth scene. One day it wasn't there, the next day it was, bands like **Alien Sex-Fiend**, **Sex Gang Children**, **Gene Loves Jezebel**, **Virgin Prunes** and **X-Mal Deutschland** springing apparently full-formed from the grave of their own musical psychoses.

Bauhaus

Any distinctions which may be drawn between the dichotomous primitive technologies towards which the latter day Punk-New Wave bands (**Pop Group**, **Slits**, **Blurt**) were moving, and the total improvisation of the early industrial acts - **Cabaret Voltaire** and **Throbbing Gristle**, can generally be bridged by the Goth example: the essentially rhythm-based dischordancy which lay at its heart; the increasing reliance on mechanical, as opposed to organic, sound effects (**Sisters of Mercy** employed a drum machine, many others utilized synthesizers); and so forth.

Nowhere is this cross-fertilization better evidenced than in the work of the brilliant **Rozz Williams**, whose three bands of greatest note - the original (pre-1983) **Christian Death**, **Premature Ejaculation** and **Shadow Project**, wilfully create archetypes for both pure Gothic and Industrial streams and coalesce them into one.

The Sisters of Mercy, contrarily, applied elements of the two regimes to their music, although it was not until 1991's Vision Thing that the awesome power fermenting behind **Andrew Eldritch**'s black-suited visage was truly unleashed; in the meantime, he had experimented with both **Meatloaf-**esque pomp (Floodland), and **Suicide-**al minimalism - the **Sisterhood**'s Gift, recorded with **Alan Vega** amongst others.

Alien Sex Fiend

(Also worthy of note is **Soft Cell**'s **Marc Almond**, whose flirtatious narcissism both drew and repelled a scene in which his chosen dress of black and bones was the height of convention. The two **Marc and the Mambas** albums are his most dramatic contribution to the Goth scene.)

If Williams, Almond and Eldritch (with a nod to Bauhaus, if only for the alienated worlds conjured up on Mask and The Sky's Gone Out) were the Goth field's sole point of original contact with Industrial music, the link would already be unimpeachable. The early 1990s, however, have also seen sterling contributions made by **Empire Hideous** and **Voice of Destruction**, neither of whom balk at further distortion of the already fuzzy barriers. (**see Premature Ejaculation, Punk Rock, Soft Cell, Suicide, Voice of Destruction**)

GROTUS

San Francisco's most important industrial band yet was formed in 1989 by vocalist **Lars Fox**, bassist **John Carson**, guitarist **Adam Tanner**, and ex-**Pagan Babies**' drummer **Bruce Boyd**.

Primarily a multi-media show, splicing cartoons, live action and music, Grotus' earliest work relied heavily on sampling; more recently, a heavier rock element (roughly comparable to **Skinny Puppy** playing "Iron Man") has crept in.

Discography: Brown (Spirit 1992); Luddite (Spirit 1993); Slow Motion Apocalypse (Alternative Tentacles 1993).

THE HAFLER TRIO

Having quit **Cabaret Voltaire** to work in television in 1981, **Chris Watson** returned to recording in 1984 with **Andy McKenzie** and **Dr Edward Moolenbeck**.

Describing their music as an extension of the Cabs' sound improvisations, the group's first album, Bang! - An Open Letter was released on the Cabs' own Doublevision label.

Two Belgium-only 12" singles followed, Alternation, Perception and Reality (1985) and the sound collages of Seven Hours Sleep (1986); 1986 also brought the interesting (but ultimately unfulfilled) The Sea Org, a 10" audio/visual package.

Watson returned to television in 1987, and with the departure of Moolenbeck, McKenzie became all three members of the Trio at once. Kill the King and the Masturbatorium EP, however, prove that the sundering was not fatal. **Clock DVA** mainspring **Adi Newton** and former porn queen **Annie Sprinkle** both guest on King.
(**see Cabaret Voltaire, Richard Kirk, Stephen Mallinder**)

Discography: Bang! An Open Letter (Doublevision 1984); Alternation, Perception & Reality (Laylah 1985); Seven Hours Sleep (Laylah 1986); The Sea Org (Doublevision 1986); Kill The King (Silent 1991); Masturbatorium EP (Silent 1991).

PETER HAMMILL see **Public Image Ltd, Punk Rock, Van Der Graaf Generator**

HAPPIEST PLACE ON EARTH, THE see **Premature Ejaculation**

HAWKWIND see **Robert Calvert, Punk Rock**

HENRY COW see **Fred Frith, Residents, Slapp Happy**

HERE AND NOW see **Punk Rock**

HOLGER HILLER see **Palais Schaumburg**

HILT

Celvin Key of **Skinny Puppy** joined forces with **Alan Nelson** for the very Puppy-like Call The Ambulance album, a storming recapitulation of Puppy's own basic intentions, set to Nelson's raucous guitar. (**see Skinny Puppy**)

Discography: Call The Ambulance Before I Hurt Myself (Nettwerk 1989); Journey To The Center Of The Bowl (Nettwerk 1991).

ROWLAND S. HOWARD see **Einsturzende Neubauten**

HUMAN LEAGUE

Dispense with **Phil Oakey**'s painful synthipop period (Dare onwards), and Human League was a very sharp stick in the hands of **Ian Marsh** and **Martin Ware**, two synthesizer buffs with more than a passing ear for what was going on elsewhere in their Sheffield, England home. Early bands - **the Dead Daughters** and, following Oakey's arrival, **Future**, were unrecorded.

"Being Boiled", this incarnation of the League's best-known song (thanks to its obtrusive presence on a post megastardom Greatest Hits album), was followed by the equally brilliant The Dignity of Labour Parts 1-4 EP, attracting the immediate attention of Virgin Records.

The League's first album, Reproduction, was highlighted by the "Empire State Human" single, a song which described, if it did not quite sound like, a building site, the album immediately established Human League in the forefront of the burgeoning New Romantic movement, a trap which Ware and Marsh seemed to fear considerably more than Oakey. (A side project single, "I Don't Depend On You", released under the alias **The Men**, catches the band in lighter mood - a harbinger of things to come!)

Travelogue, in 1980, finds the group already splitting, with Marsh and Ware now looking towards the direction they would take with their own **British Electric Foundation/Heaven 17** project, and doing their best to pull Oakey along with them. He ain't listening, though, and the resultant schizophrenia makes for a considerably more uncomfortable album than anybody, the band and their record company included, expected.

Separately, both Human League and Heaven 17 now settled down to make considerably shinier records than either would have dreamed of during their days of early experimentation; both similarly attempted to escape from them, Oakey through a 1985 collaboration with producer **Giorgio Moroder**, Ware/Marsh through the occasionally eclectic **BEF**. (**see Synthipop**)

Selected discography: Dignity of Labour Parts 1-4 (Fast 1979); Reproduction (Virgin 1979); Travelogue (Virgin 1980).

IMAGES IN VOGUE

Formed in Vancouver in the early 1980s, IIV - like the similarly intentioned **Moev** - performed regularly at the Luv-a-Fair gay bar, at that time the city's own outlet for electronic music.

IIV released a number of indie singles locally, produced by **David Ogilvie**, before signing to WEA and landing tours with **Roxy Music** and **Duran Duran**. Sadly, says Gordon, this marked "the beginning of the end." Running into "a wall of general disinterest/ manipulation" from their record company, IIV suffered several personnel losses (Gordon quit immediately prior to the band's debut album), and with their sound now a pale reflection of its original majesty, the group shattered around 1985.

Members of IIV went on to form **Numb** (**Don Gordon**) and **Skinny Puppy** (**Kevin Crompton**); in addition, the group's monitor technician, **Ric Arboit**, joined with members of Moev to found Nettwerk Records, and soundman **Greg Reely** engineers for Skinny Puppy. **see Moev, Numb, Skinny Puppy**

Selected discography: Educated Man EP (indie 1981); Lust for Love EP (WEA 1982).

In The Nursery

IMMACULATE CONSUMPTIVES see Foetus Inc, Soft Cell

IN THE NURSERY

Klive and **Nigel Humberstone** strive to create modern classical music in an electronic vein, an area otherwise best explored by **Laibach**. Although their albums show little progression, each is recommended.

Discography: L'Esprit (Third Mind 1990); Sense (Third Mind 1991); Duality (Third Mind 1992).

INDUSTRIAL ARTZ

Chicago-based band which melds industrial, rap and metal influences into a wicked whole, best exemplified by a version of "Psychotic Reaction" included on their debut. **Led Zeppelin**'s "Kashmir" is a surprise sample on "Power Trip".

Discography: 6 Demon Bang (Cargo 1991).

INDUSTRIAL MUSIC

"There's an irony in the word 'industrial' because there's the music industry. And then there's the joke we often used to make in interviews about churning out records like motorcars - that sense of industrial. Up until then the music had been...based on the blues and slavery, and we thoght it was time to update it to at least Vctorian times - the Industrial Revolution." - **Genesis P-Orridge** 1983

It was **Monte Cazazza**, a San Francisco artist, who coined the phrase "Industrial music for industrial people", but **Throbbing Gristle** which put a sound to his ideas. "When we finished [our] first record," P-Orridge said, "we went outside and we suddenly heard trains going past and little workshops under the railway arches and the lathes going and electric saws and we suddenly thought, 'we haven't actually created anything at all, we've just taken it in subconsciously and re-created it."

The theory behind the music was based around taking risks. "Sometimes," P-Orridge once mused, "I think we'e given birth to a monster, uncontrollable, thrashing, spewing forth mentions of Auschwitz for no reason. It's funny, because when I really think about it, the original half-dozen who started it all off are still the best ones, like [**Cabaret Voltaire**], **Boyd Rice**... the first wave. But I suppose that's inevitable, it's the old story, like in the 60s - **Zappa** was completely different to **Beefheart** who was completely different to the **Doors** who were completely different to the **Velvet Underground**. And we were completely different to the Cabs, and Boyd was totally different and **Z'ev** is different. We've all got a quite clear individual style linked to our individual lives whereas now they all sound like each other, and more than each other they sound like weedy fragments that they've honed in on, of one or the other people

"The staggering thing [in September, 1982], is that 'industrial' has become a word that is used worldwide - there are record sections in shops in Japan that say 'Industrial Music'. Journalists now use 'industrial' as a term like they would 'blues'. It's become part of the vocabulary. And I'm sure most of them have totally forgotten where it came from!"

The arrival on British shores of the German group **Einsturzende Neubauten** gave the scene fresh impetus; **DAF**, and a resurgent interest in **SPK** maintained the momentum, but it was **Depeche Mode** who grasped the bull by the horns and pulled it into the full commercial spotlight. The impact of their actions has been nullified somewhat by time (and a few water-treading records), but Depeche remain the catalyst around which many of the bands from the mid-1980s and beyond were created. (Plus, it is no longer fashionable to claim Depeche as an influence!)

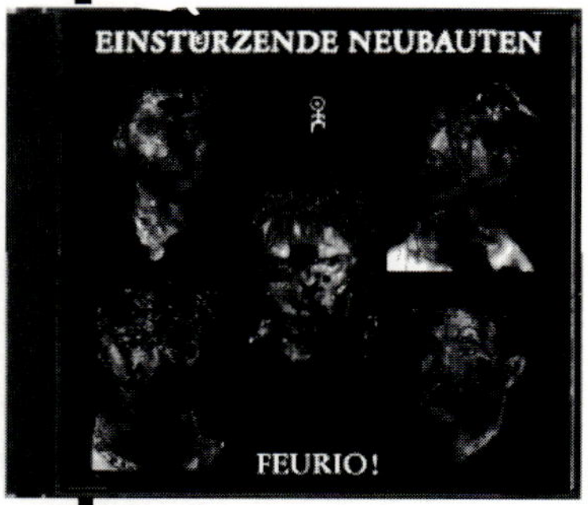

With its own natural boundaries becoming increasingly blurred, until even pure ambient, spontaneous and improvisational music forms were considered as valid a point of reference as any, the industrial scene now moved in on sundry unrelated areas of music - most notably the modern dance and decades-old Progressive Rock scenes, imbibing from these much of the energy needed to sustain itself. Surprisingly, these hijacked elements of classic **Hawkwind, King Crimson**, Captain Beefheart and Frank Zappa seemed right at home alongside hip hop and computer technologies, and the massed advance of sampling opened up even wider areas of influence and expression. Today, even abrasiveness and anger are no longer pre-requisites within the field, a reminder, possibly, that even machines can be quiet sometimes!

Nevertheless, the true nature of industrial music remains questionable; like punk in the late 1970s, it embraces so many divergent streams of thought that there is no longer a definable "industrial" sound - assuming there ever was.

"People take the label to describe music," says **Cosey Fan Tutti**. "Industrial for [Throbbing Gristle] embraced everything we did, every activity. It wasn't just the music, it was a philosophy. It was a seriousness of what life is about. That has nothing to do with what is called industrial now. It was so anti-music to call something industrial."

And the future? According to **Chris Carter**, there is none. "Industrial music stopped when Throbbing Gristle and Industrial Records packed up. And there's been nothing truly Industrial since then. There aren't any true Industrial bands."

INNER SPACE see **Can, Holger Czukay, Kraut Rock**

INTERMIX see **FRONT LINE ASSEMBLY**

JAPAN see **Holger Czukay, Synthipop**

JAYWALKER
A Swiss band whose second album, I Told Them My Dreams (Exciter Records) was released with the warning, "experimental trash industrial hardcore."

JOINED AT THE HEAD see **Controlled Bleeding**

JON JON
Early 80s techno-industrialists Jon Jon was formed by **Lewis Cypher**, as a combination of the technical essence of **Kraftwerk** and the aggression of **Einsturzende Neubauten**.

Despite attaining a certain popularity in Germany, the group meant little in the US and split in 1988. Cypher went on to form **Voice of Destruction** **(see Voice of Destruction)**.

JOY DIVISION

"To a large extent, [Joy Division] evolved a unique sound of their own, a new sound amongst the sort-of progression of musical sound. They had their own, recognizable, individual style. And then suddenly there's 50 other groups who, because they like Joy Division, use the same style [but] can't have the content because they're not **Ian Curtis**" - **Genesis P-Orridge** 1982

Arguably, without Ian Curtis' tragic, but in terms of rock legend so romantic, suicide in May 1980, Joy Division would today be little more than a passing notation in history, a band whose posthumous reputation certainly rises high above their actual capabilities.

Having been in the future-star-studded audience when the **Sex Pistols** played Manchester in 1976, **Peter Hook** and **Bernard Sumner** **(Albrecht)** were joined by Curtis and drummer **John Tobac**, the latter arriving just days before the **Stiff Kittens** (as **Buzzcock Pete Shelley** suggested they be called) became **Warsaw**, itself an event which occurred on the eve of their May 1977 live debut at the Electric Circus.

Joy Division

The band owed little to the then-prevalent local scene, riding high as it was on the pure pop sensibilities of **Slaughter & The Dogs** and **Buzzcocks**, and the zanier, but still conventional **Drones** and **Nosebleeds**. Nevertheless they persevered, replacing Tobac with first, **Steve Brotherhood** and a month later, **Steve Morris**. (According to legend when Brotherhood quit to join **Panik** he tried to take Curtis with him.)

The emergence of **Howard Devoto**'s **Magazine** in January, 1978, however, exerted considerable influence and, perhaps, pressure upon a foursome whose only vinyl to date, the Ideal For Living EP, had shown them to be more in tune with mutant **Bowie** than the power of punk.

Injecting into their sound the ragged despair which they later all but patented, Warsaw was close to signing to RCA when the local Factory label swooped - the eleven tracks recorded for a projected album, in May 1978, appeared on bootleg in 1980, and show a heavy debt to Devoto and co, most noticeably in the four tracks rerecorded from the original EP - "Warsaw", "No Love Lost", "Leaders of Men" and "Failures". Another two songs from this session, "Ice Age" and "Walked In Line", were rerecorded the following year and appear on the Still retrospective.

Joy Division's next releases were two tracks on the seminal A Factory Sampler EP ("Digital" and "Glass"), and two more on the Fast label's Earcom 2 EP ("Auto-Suggestion" and "From Safety to Where"); they can also be heard live on the Short Circuit at the Electric Circus compilation ("At A Later Date").

Unknown Pleasures, the band's debut album, contains by far their finest recordings: "She's Lost Control", "Shadowplay" and "Disorder"; absent from the set, despite being included in the band's first John Peel Session was the single "Transmission", another essential JD piece.

Also of interest is the French single of "Atmosphere"/"Dead Souls"; this was reissued following Curtis' death with a new version of "She's Lost Control", widely believed to have been taped in March, 1980.

New Order

Championed by a small, but vociferous section of the UK press, Joy Division spent much of late 1979 touring, including a lengthy UK outing supporting Buzzcocks. The band was just preparing to release Closer when Curtis died.

A denser, darker, but ultimately less-satisfying record than its predecessor; the demons which were gathering around Curtis were present, but still unfocussed, a sense of personal alienation which Joy Division were able to capture only in small bursts - "Isolation", from Closer, is perhaps the ultimate Joy Division track.

Curtis' death, at a time when Joy Division were still little more than a respected minor cult, was not, however, to spell the end of the band, and was in fact wholly responsible for their later emergence. A single, the desolate "Love Will Tear Us Apart", shot the group into the UK chart, although arguably it remains the ugliest song Joy Division released.

Thereafter, the Joy Division discography goes into overdrive, resurrecting material of such often dubious quality that it is surprising the group's reputation ever emerged intact. The circus opened with the flexidisc release of "Komakino", "Incubation" and a third, untitled instrumental drawn from the Closer sessions. Distributed free of charge to record stores, the two titled tracks later reappeared on the CD compilation Substance.

This was followed by Still, a retrospective divided between studio and live material; added poignancy was added to the set when it was revealed that the concert, in Birmingham, was in fact Curtis' last. That did not, however, excuse the general stodginess of the album, and claims that this was simply a "bad night" were not helped by another of Still's live cuts, Joy Division haplessly flailing around the **Velvet Underground**'s "Sister Ray". (Live footage accumulated on the Here Are The Young Men video is similarly lackluster.) For anybody wishing to pursue the Joy Division legacy further, the live disc also includes an embryonic "Ceremony", later to become **New Order's** debut single.

The most important Joy Division exhumations were the band's two John Peel sessions, from January and November, 1979. The second, incidentally, includes a rendition of "Love Will Tear Us Apart" several light years more accomplished than the hit version. Later collected onto one album, these recordings, plus the Substance compilation, should satisfy most newcomers' cravings for a taste of the band.

Selected discography: An Ideal For Living EP (Enigma); Unknown Pleasures (Factory 1979); Closer (Factory 1980); Still (Factory 1981); Substance (Qwest 1988); The Peel Sessions (Strange Fruit 1990).

JUSTIFIED ANCIENTS OF MU MU see KLF

HENRY KAISER see Fred Frith

KILLING JOKE

Erupting out of London's post-punk apocalypso scene, drawing in equal parts from the vicious mindfuck of **Birthday Party** and the more ponderous savagery of **the Pop Group**, Killing Joke - **Jaz Coleman** (vocals, keyboards), **Martin "Youth" Glover** (bass), **Geordie** (guitar) and **Paul Ferguson** (drums) - lurched into play with Almost Red, a disquieting four song EP in which music and melody were treated as two very separate notions, and scarcely allowed to touch.

An immediate underground success, the group signed to EG, home of **Roxy Music** and **King Crimson** (their two most obvious, if barely recognizable, antecedents), and released Killing Joke, a statement of intent which basically left the listener with two choices. Either the world was ending - or he was.

Noting the growing infusion of funk into the new wave's lexicon, What's THIS For was a considerably easier record, echoing in equal parts the seminal **Gang of Four** and, through the reggae-tinged ghosts which hung around the fringes, a controlled version of **the Slits**. But such reference points remain vague at best; creating ice sculptures of chaos, Killing Joke remained unique.

Produced by **Connie Plank**, the band's third album, Revelations, marked both the high- and low-water mark in their career. Musically, it exceeds even the peaks to which Killing Joke had customarily climbed; however, it also precipitated a press backlash from which the band barely recovered.

Coleman and Geordie quit, heading off to Iceland for a brief sabbatical. Their return to the band later that same year heralded a vivid change in direction; with Youth having been replaced by **Paul Raven**, Killing Joke dumped much of their earlier bleakness in favor of a dancey playfulness amidst which their stil rampant dark humor often seemed ill-at-ease. (Youth moved into production, working with **Executive Slacks**, amongst others.)

To these ends, Killing Joke rushed out an album, an EP and the exhilarating live Ha! EP in less than a year. Fire Dances, however, was to be the last group effort in two years; when they returned in 1987, Killing Joke had undergone even further musical refinement. Night Time contains the UK hit single "Love Like Blood", and even the band's most fervent admirers found themselves describing Night Time in terms they might never have dreamed of applying to Killing Joke - catchy, danceab e, commercial.... (All the band's albums to this point were later repackaged in the boxed set An Incomplete Collection.)

Another year-plus break was in order, but Brighter Than A Thousand Suns only hastened Killing Joke's decline. With synthesizers now all but replacing guitar, the last traces of the group's early adventure and outrage were now on the chopping block. Outside The Gate was the ax which beheaded it - and the band. Killing Joke broke up in 1988.

But they wouldn't stay dead. With former **PiL** drummer **Martin Atkins** replacing Paul Ferguson, and the Coleman-Geordie axis, the band were determined to push Killing Joke back into the experimental fore they had once so nearly dominated. The reconstituted Joke released Extremities, Dirt & Various Repressed Emotions, the first album in eight years to truly deserve the band's name and reputation.

The band's ultimate, and severely acrimonious, demise in 1991 was followed by the Coleman-less survivors forming **Murder Inc.** under Atkins' aegis. (**see Martin Atkins, Executive Slacks, Pigface, Public Image Ltd**)

Discography: Almost Red EP (Malicious Damage 1979); Killing Joke (EG 1980); What's THIS For? (EG 1981); Revelations (EG 1982); Birds of a Feather EP (EG 1982); Ha! EP (EG 1982); Fire Dances (EG 1983); Night Time (EG 1985); Brighter Than A Thousand Suns (EG 1986); Outside The Gate (EG 1988); The Courtauld Talks (**Coleman/Geordie** spoken word Invisible 1989); An Incomplete Collection (EG 1990); Extremities, Dirt and Various Repressed Emotions (Noise International 1990).

KING CRIMSON see Robert Fripp

RICHARD H. KIRK

Throughout **Cabaret Voltaire**'s period of greatest success, founder member Kirk has maintained a constant stream of solo work, beginning with the cassette only Disposable Half-Truths. Following very much in the vein of the band's original work, and thus silently condemning the increasingly commercial sounds which he and **Stephen Mallinder** were making elsewhere, both Disposable Half Truths and 1983's compilation of three years of solo doodlings, Time High Friction (Doublevision) were as powerful as anything Cabaret Voltaire had done in the past.

1985 brought Leather Hands (Douvblevision), a collaboration with **Box** lyricist **Peter Hope**; and the following year, Kirk returned to the Cabs' old label, Rough Trade, for a flurry of new releases - the brilliant "Hipnotic" 12" single, the album Black Jesus' Face and a mini-LP, Ugly Spirit. There was also a second collaboration with Hope, the Hoodoo Talk album (Internal).

Late in 1989, Kirk began working in the field of so-called Bleep music, teaming up with Sheffield **DJ Parrot** under the name **Sweet Exorcist**. Test One was a single dance track released in six different mixes over the course of two 12" EPs (Warp). A second Sweet Exorcist project produced the Clonk Remix EP at the end of the year, and in January 1991, the album Clonk's Coming (both Warp). Clonk, incidentally, is a character from the 2000 AD comic book.

Kirk is also one half of **Xon**, with **Robert Gordon**. Their Mood Set EP was released in early 1991 (Network), and with fellow Cab Steve Mallinder, a member of the **Ministry-**fired **Acid Horse.** (**see Cabaret Voltaire, Hafler Trio, Stephen Mallinder, Ministry**)

Discography: Disposable Half-truths (Industrial 1981); Time High Friction (Doublevision 1983); Leather Hands (Douvblevision 1985); Black Jesus' Face (Rough Trade 1986); Ugly Spirit (Rough Trade 1986); Hoodoo Talk album (Internal 1986); Test One (with Sweet Exorcist Warp 1989); Clonk Remix EP (with Sweet Exorcist Warp 1989); Clonk's Coming (Warp 1990); Mood Set (with Xon Network 1991).

KLF

The hip-hop house acid rave thrash supergroup, KLF - also trading under the names **Disco 2000, Space, the Orb** and **the Justified Ancients of Mu Mu** - is the brainchild of former Zoo Records' boss **Bill Drummond** and former **Zodiac Mindwarp** guitarist **Jimmy Cauty**. KLF stands for Kopyright Liberation Front, and was essentially designed to test Britain's rigorous copyright laws - could sampling, in more or less absurdly obvious chunks, hold up as art in a court of law?

Swedish supergroup **Abba**, one of several major league victims of the Justified Ancients' 1987 (What The Fuck's Going On) album, decided it couldn't and launched a law suit which effectively killed the album dead. The publicity, however, enabled the duo to continue, reissuing 1987 as The JAMS 45 Edits EP, from which all unauthorized borrowings had been removed. The result was somewhat akin to a very funny joke being retold without its punchline, and the following year the JAMS tried again, this time asking permission before creating the merry pop collage.

Who Killed The JAMS? prefaced the release of the duo's greatest hit, a monstrous arrangement of **Gary Glitter**'s "Rock'n'Roll" thematically reworked around the television program Dr Who. Released in the name of **The Timelords**, "Doctoring the Tardis" became a UK chart topper during the summer of 1988, and three years later remains the Drummond-Cauty team's finest moment... not to mention the last great #1 in British pop history. The Doctoring The Tardis CD EP, incidentally, includes a priceless contribution from Glitter himself, "Gary Joins The JAMS".

"Tardis" reappears on the History of the JAMS album, a compilation which hauls in three earlier singles (themselves already compiled onto the Shag Times EP, and worth hearing for "**Whitney (Houston)** Joins The JAMS" alone) plus "The Porpoise Song", in which disco queen **Donna Summer** is given the JAM treatment.

The JAMS passed away at this point, Drummond and Cauty now reassigning the KLF label to their own music (hitherto it had been their record label!) and, retaining only the love of irreverent sampling which had made the JAMS so special, set off down a somnambulistic avenue which adds little to the sub-Eno ambience they have most obviously borrowed from beyond the sound-bytes they layer over them. Cauty's **Orb** psuedonym takes this to even greater extremes, often dispensing with the sound-bytes as well. A waste of electricity.

Selected discography: 1987 (What The Fuck's Going ON?) (JAMS KLF 1987); 1987 (The JAMS 45 Edits EP) (JAMS KLF 1988); Who Killed The JAMS? (JAMS KLF 1988); Doctoring The Tardis EP (Timelords TVT 1988); Shag Times EP (JAMS KLF 1989); The History of the JAMS (JAMS TVT 1989).

KLUTE

Claus Larsen, the Danish brain behind Leaether Strip, inaugurated Klute in 1990, debuting the project the following spring with the Explicit 12".

Considerably harsher than **Leæther Strip** – or most other industrialists for that matter – (Industrial Nation claims Klute makes "**KMFDM** and **RevCo** sound like elevator music"). Klute material began turning up in **Leæther Strip's** own live act, and in 1992 the first album was released to wide acclaim and worldwide licensing deals. The US edition is being handled by Cleopatra Records. (see **Leæther Strip**).

Klute

Discography: Explicit EP (Zoth Ommog 1991); Excluded (Cleopatra 1993).

KMFDM

Stepping out from where the much-missed **Belfegore** left off, this multi-talented Hamburg trio (whose moniker has been translated as everything from Kill Mother Fucking **Depeche Mode** to Kein Mitleid Fur Die Mehrheit - no pity for the majority) first came to note with a string of singles for the German Skysaw label, released between 1984-86. Gathered together on the What Do You Know, Deutschland album, they indicated a band whose industrial roots were tangled amidst a variety of similarly noisy, but often quite disparate musical themes - Heavy Metal thrash, and sundry dance routines.

Signing to Chicago's Wax Trax label, KMFDM linked with pioneering producer **Adrian Sherwood** in 1988 to cut Don't Blow Your Top, a spartan album which, like most of Sherwood's work, highlighted the producer more than the band. The following year's UAIOE is a considerably more representative album, and was followed by the brilliant "Virus" and "Godlike" singles, the latter of which was revamped for Naive. (Another 1990 single was split with future cult stars **My Life With The Thrill Kill Kult**).

KMFDM

Money, an album upon which vocalist **Sasha Konietzko**'s vocals finally moved into the **Andrew Eldritch** territory they had frequently threatened to invade, preceded a successful autumn 1992 US tour (as a busy sextet), and while the set remained heavily industrial, it was within a jarring pop framework which, with a little more refinement, could see KMFDM become the scene's first true superstars.

Excessive Force is a side project by **Sasha Konietzo** and Thrill Kill Kult's **Buzz McCoy. (see My Life With The Thrill Kill Kult, Pigface)**

Discography: What Do You Know, Deutschland (Skysaw 1986); Don't Blow Your Top (Wax Trax 1988); UAIOE Wax Trax 1989); Naive (Wax Trax 1990); Money (Wax Trax 1992); Conquer Your World (Excessive Force Wax Trax 1992).

KMFDM

KRAFTWERK

In 1991 Kraftwerk released The Mix. **Ralf Hutter** described it as "a retrospective that goes forward", which is a neat way of saying it's a compilation remix which swoops down upon eleven of the German band's best-loved songs, turning this bit up and that bit down and occasionally throwing everything up into the air to see which order it landed in.

As a rescheduling of a grand plan, however, The Mix added nothing to the inherent majesty of Kraftwerk's work, altering texture which had been perfect to begin with, and proving that it was not technology which Kraftwerk relied upon, but the ability to use that technology to its greatest extent. Updating the technology alone could never work because Kraftwerk depended on the inspiration of the moment, not that of twenty years after the moment has passed. One to avoid.

Kraftwerk

It is, however, the only Kraftwerk album which shculd be avoided. Although it was 1974 before Kraftwerk came to international attention, gatecrashing both the British and American charts with a mercilessly edited rendering of their fifth album's 22 minute title track, for three years the quartet had been quietly etching themselves a reputation which even today, still echoes through the alternative rock scene. Other names may have made a greater splash in both musical and commercial terms, but when debating the single most influential band of the 1970s, Kraftwerk wins hancs down.

Hutter and **Florian Schneider-Esleben** first worked within Germany's freeform jazz scene, not moving into electronics until they witnessed **Terry Riley**'s two-week long performanmce of sine wave oscillators at a Munich art gallery in 1969. Immediately inspired to take Riley's work to extremes, they joined with **Butch Hauf**, **Fred Monicks** and **Basil Hammoudi** in **Organization**, whose four track album Tone Float, was released in 1970.

Produced by **Connie Plank**, Tone Float was most heavily influenced by **Pink Floyd**'s contemporaneous dabblings in fragmented sound structures (the solo pieces on Ummagumma), and clearly left Hutter and Schneider unhappy. Within the year they had left Organization, and formed Kraftwerk with **Klaus Dinger** and **Thomas Homann**.

Kraftwerk, arriving in a sleeve depicting a power plant (the band's name translates as such), laid much of the ground from which the group would in later years take off; when Kraftwerk and its successor, Kraftwerk II, appeared in the UK as a double album, the promotional campaign described it as "a new experience in music", a vague but understandable term. There simply were no reference points for the repetitive, mechanical hypnotics in which Kraftwerk specialized.

Dinger and Homann had quit to form **Neu** by the time work began on Kraftwerk II and Kraftwerk remained a duo for their next album, Ralf and Florian. By late 1973, however, guitarist **Klaus Roeder** and percussionist **Wolfgang Flur** had been inducted and work began on what was to become the most important album in the history of electronic music, Autobahn.

What inspired the group's UK and US labels to harshly edit the 22.30 minute title track down to single length remains one of the greatest mysteries in rock'n'roll. Certainly this electronic recounting of a drive down one of Germany's open freeways had nothing in common with anything else vying for chart glory - 1974 was the year, in Britain, of Glam Rock and Disco; in the US, the likes of **Terry Jacks** and **Paper Lace** epitomized what the pop kids wanted.

But Autobahn, half novelty, half incomprehensible weirdness, made a mockery of them all, and with the album making similar commercial headway, Kraftwerk was poised for a major breakthrough. They even provided a welcome distraction within the increasingly cloistered disco scene, although it is hard to see today how they fit in alongside **Barry White** and the **Hues Corporation**. The only sour note in all of this was the band's failure to make any money from their success - they had sold the rights of Autobahn to their record company for a mere $4,000.

Roeder quit the band in October 1975, while the remaining trio were putting together first, their own Kling Klang label, and then adding the finishing touches to an arsenal of electronic instruments which they saw as replacing the still-conventional guitars, drums and unweildy synthesizers which they were then carrying around. Radioactivity, their next album, was very much a testing ground for this impressive array of tricks, and though the album in no way lived up to the success of its predecessor, it showed that success had in no way gone to Kraftwerk's head.

Interest in Kraftwerk remained high, of course, and 1976 saw a slew of compilation albums appear throughout Europe, drawing in the main from the band's first three albums. Of these, the UK Exceller 8 and the French Kometenmelodie: A Tout Collection are both recommended, individually if not collectively. A later compilation, Elektro Kinetik, expands the data base to include selections from "Autobahn".

Kraftwerk spent much of that year refining further the progress made on Radioactivity, but arguably the biggest break in their career so far came when producer **Giorgio Moroder** took American **Donna Summer** into the studio to record "I Feel Love", a disco song on which Summer's vocals soared above an array of dense, throbbing syntheizers - the motorik beat which **David Bowie** and **Eno** had recently applied to much of **Iggy Pop**'s The Idiot comeback album.

Bowie, Pop and Eno, too, loomed large in Kraftwerk's fortunes, dropping their name as an influence on their own work (Bowie's Low, recorded like The Idiot in Berlin, borrowed much from Kraftwerk's earlier experiments in sound texture). Into this climate of anticipation, Kraftwerk released Trans-Europe Express (and repaid the favor by namechecking Bowie and Iggy in the title track; a debt Bowie reactivated on his Heroes album, with "V2 Schneider").

Perhaps the band's most fulfilled vision, despite the title track merely rescheduling "Autobahn" down a railroad track, Trans-Europe Express joined the Bowie-Eno-Moroder axis in redefining the direction of the now-burgeoning New Wave. Everybody, it seemed, took a leaf from its book - the savage automation of "Showroom Dummies" provided **Gary Numan** with a role model which would ultimately take him to #1; the clanging "Metal On Metal" brought into harsh realism the possibilities of employing "found" instruments - in this case sheets of metal, hammers and the like - in a pop context, the almost transcendental "Europe Endless" provided the framework around which a number of lesser groups (the memorable **Rikki & The Last Days of Earth** included) would base their own excursions.

But Kraftwerk were by no means finished. The Man Machine, the following year, made even "Trans Europe Express" seem somehow old-fashioned. Rooted in an exploration of man's relationship with machinery, and the possibilities that a blending of the two might imbibe, Kraftwerk's seventh album leaped into the UK Top 10, trailing in its wake two essential singles, "The Robots" and "Neon Lights".

Kraftwerk also took this opportunity to debut their own automatic doppelgangers at a pair of press-only shows in Paris and New York. This ultimate display of the machine's precedence over man was at the same time heroic and foolhardy - critics, already aiming ever-sharper barbs at Kraftwerk's musical inhumanity, now had a field day when considering the potential of the robots to completely replace mankind in music. The robots were also on hand for press conferences and other personal appearances, with Kraftwerk admitting that if they could, they would happily stay at home and let the automaton do all the work for them. Their critics went into overdrive.

Kraftwerk, of course, were unperturbed. The three year sabbatical they now embarked upon was for technical reasons only - they were re-equipping their studios with computers. And when they did re-emerge, with the Computer World album, it was to discover that their detractors had been trampled to death. Britain was now in the grip of the New Romantics movement, a sallow, but, in essence, musically challenging movement, in which synthesizers and electronics had all but wiped guitars and drums from the face of the planet. **Gary Numan, Ultravox, Orchestral Manouvres In The Dark, A Flock of Seagulls** and so on now ruled the roost, and even as Computer World moved up the chart, so the band's UK record company began reactivating Kraftwerk's back catalog.

Trans Europe Express reappeared in the album chart, but more telling was the band's singles success. From the new album, "Pocket Calculator", performed in England on the A-side, Japanese on the b-side ("Dentaku") made #39; "Computer Love" #36. Then someone decided to flip the record over, and "The Model" (from Man Machine) soared to #1, even as "Showroom Dummies" was moving up to #25. Suddenly Kraftwerk - the first German band ever to top the British chart - was the biggest thing in sight. Had they followed it up, they might have remained so.

Instead they went to earth and it was 18 months before a new Kraftwerk album was even rumored. A 12" single, "Tour de France" (performed in French and German; no English language version ever materialized), appeared on the continent and was imported heavily into the UK, where EMI was still awaiting news of a new album.

Finally giving in to demand, the single was given a domestic release alongside the news that a new album, Techno Pop, was forthcoming. Comprising just four tracks ("Tour de France", "Sex Object", "The Telephone Call" and the title track), the album had already reached mix-down stage when Kraftwerk suddenly decided to abort the project. Three more years were to pass before the three new songs finally made an appearance, on the Electric Cafe album, which paired Kraftwerk with producer Francis Kervorkian.

It was not a smart move. Talented though Kervorkian was, his vision frequently clashed with Kraftwerk's own, giving the impression that he had been called in merely to salvage something from a project which was in danger of collapsing beneath its own weight. Electric Cafe remains the least important Kraftwerk album.

The band remained in hiding throughout the remainder of the 1980s, resurfacing only in a brief flurry of activity around the time of The Mix. A stronger retrospective, The Model, was released by Cleopatra shortly after.

The eternal "Autobahn" notwithstanding, Kraftwerk's period of greatest influence and success was unquestionably the four year period bookended by "Radioactivity" and "The Man Machine", during which even the self-conscious namedropping of Bowie and Iggy seemed somehow innovative; but the clattering ironworks of "Metal on Metal" a bizarre Teutonic joke. Whoever would have thought an entire musical genre could be spawned by such tomfoolery?

Kraftwerk

The Model concentrates on the band's British and US singles career, with the added bonus of the French-language "Showroom Dummies", "Les Mannequins". (All Kraftwerk's material was also available, in their homeland, in German) The essential continuity of the pivotal Trans-Europe Express, however, is preserved; "Express" itself segues immediately into "Metal On Metal", after which "The Robots" - Kraftwerk's most condescending nod to their own reputation - comes almost as a relief!

The release of The Mix prompted a great deal of celebration amongst Kraftwerk's still loyal following, offering as it did remixes of several of the cuts included on The Model. The party proved premature - Kraftwerk's strength was their simplicity, and their understanding of the technology with which they worked at their peak. Attempts to meddle with the sound simply muddled the signals they were sending in the first place. The Model, then, restored dignity to Kraftwerk's back catalog...and reminds us of just how different the last decade would have been without them. **(see Kraut Rock, Neu, Punk Rock, Terry Riley, Donna Summer, Synthipop)**

Discography: Tone Float (as **Organization** RCA 1970); Kraftwerk (Phillips 1970); Kraftwerk II (Phillips 1972); Ralf & Florian (Phillips 1973); Autobahn (Vertigo 1974); Excellor 8 (Vertigo 1975); Radioactivity (Cleopatra 1975); Trans Europe Express (Cleopatra 1977); Man Machine (Cleopatra 1978); Elektro Kinetic (Vertigo 1981); Computer World (EMI 1981); Tour de France EP (EMI 1983); Electric Cafe (EMI 1986); The Mix (EMI 1991); The Model (Cleopatra 1992).

DAGMAR KRAUSE see Fred Frith, Slapp Happy

KRAUTROCK

A general, and - though it may seem otherwise - genuinely affectionate term coined to describe the insurgence of German bands of the early 1970s.

Spawned by the still fermenting spirit of cultural dischord which, in the UK and US, had been distilled and bottled by the mass media, and the belief that Woodstock (as a symbol as opposed to an event) had been merely the starting point of the true revolution, these fledgling bands reacted also to the almost absolute absence of "live" music audiences, and turned their attentions instead towards fiercely introspective, or at least contemplative, themes.

Although the early work of **Soft Machine** and **Pink Floyd** within the rock spectrum, **LaMonte Young**, **Stockhausen** and **Terry Riley** beyond it, undoubtedly played some part, bands such as **Faust**, **Amon Duul** and the splinter **Amon Duul II**, **Organization** (from whence sprung **Kraftwerk** and indirectly, **Neu**), **Inner Space** (the embryonic **Can**), **Cluster**, and **Tangerine Dream** developed in an almost absolute vacuum, reacting only off one another.

The synthesis within much of this was producer **Connie Plank**, whose early work with **Kraftwerk** and **Neu** established him as the premier German producer. His taste for icy-cold precision is not always perfect for every occasion (**Ultravox** were certainly not helped when they recorded Systems of Romance at his Cologne studio), but if there was any one single catalyst in the entire Kraut Rock field, it was Plank.

(In later years, of course, **Giorgio Moroder**'s Musicland studios vyed, with varying degrees of success, for superiority with Plank. **Donna Summer**'s "I Feel Love" and **Sparks**' Number One Song in Heaven are most representative of his skills; other work, with **Phil Oakey** and pomp rockers **Queen** leaves much to be desired.)

The escalating interest shown in their work by English-speaking audiences to some degree halted the development of the scene; by 1973 or so, Kraut Rock as a definable genre was all but redundant. Those bands which survived this semi-cataclysmic sundering (primarily Kraftwerk and Tangerine Dream) by necessity developed in very different directions.

Others, attempting to hang onto their earlier sensibilities at the same time as retaining their appeal to the mass market, ultimately fell by the wayside. The sad fate of Can, transforming themselves into a Top 40 electropop band five years before the synthiboom they anticipated, remains an object lesson to all. (**see also Amon Düül II, Can, Faust, Kraftwerk, Neu, Terry Riley, Tangerine Dream, LaMonte Young**)

LAB REPORT

Matthew Schulz of Pigface fame, spins off with Eric Pounder, Chris Blazen and Becky Allen for the largely percussive Lab Report. Repetitive rather than required listening, Figure X-71 is an hour of crunching improvisation which scarcely bodes well for future excursions. (see Pigface).

Discography: Figure X-71 (Invisible 1992).

LAIBACH

Springing from the loins of Slovenian art collective **NSK**, Laibach is the musical expression of a seemingly Teutonic urge to establish the Slovenian capital of Ljubljana as the Athens of the modern age, a goal which does make more sense when viewed in the totality of NSK's full range. The political implications of NASK's existence are best grasped through the book Neue Slowenische Kunst, published in 1992 by Amok Books.

It is through their verbal interpretations of NSK that Laibach have done much to alienate western supporters (Europe's obsession with them as a neo-Nazi splinter group, is a shadow they may never escape from); it might also be pointed out that in musical, as in artistic terms, the overt usage of society's most powerful icons will sooner or later inspire unpleasant comparisons.

Laibach

Art this intense, the argument goes, has to be unsavory, and the fact that NSK's favored symbols - a combination of pagan, Christian and, completing the circle, Communist images - are in themselves weighted down by conflicting emotions only reinforces the discomfort. Translating those same emotions to what is essentially rock and roll is simply asking for trouble.

Yet the only way to beat a tabboo is to break it, and if Laibach and NSK really are walking where others fear to tread, remember this. It's when the fear disappears that we should really start worrying. Until then we still have a chance.

Laibach themselves commenced operations in June, 1980, and played their first (banned) show at the Trbovlje Workers Center. Subsequent shows blended Laibach's music with other NSK exhibitions, but, in July 1982, the band made its first tentative steps on its own, recording four songs "Death For Death", "State", Triumph of Will" and "Jaruzelsky". The following year, they contributed to a cassette pairing Laibach with **The Last Few Days**.

In May, 1983, the band signed to RTV Ljubjana and recorded the unissued The Sky Glows album. Shortly after they linked with the Belgian Laylah label for the "Sila"/"Boji"/"Brat Moj" 12" EP. From this same period, Laibach's version of the Yugoslavian national anthem appears on an Italian compilation, World National Anthems, and the "Panorama"/"Dekret" single was released in Britain.

Laibach

Many of the band's earliest recordings, which fused jazz, rock and disco into a strangely portentous (some say Wagnerian) whole, are collected together on the Rekapitulation double album.

A European tour followed, from which the Occupied Europe, Vstajenie v Berlinu, Life in Hell, Ein Schauspieler and Neue Konservativ albums emerged during 1984-1985. Another live album, Divergences/Divisions, dates from a 1986 show in Bordeaux. Many of these, incidentally, were released on cassette only.

The band's UK profile, meanwhile, was maintained by a flurry of 12" singles on Cherry Red ("Der Liebe Ist Die Grosste Kraft") and Mute ("Geburt Einer Nation", "Life is Life") and two albums: Nova Akropola and Opus Dei. Of these, the former's use of Marshall Tito's speeches as lyrics for two of the songs prompted considerable controversy amongst critics unable to see past the band's iron cast grimaces - like **Kraftwerk**'s experiments with robots, much of Laibach's work can be considered ironical, or at worst, allegorical, but is hampered primarily by their audience's refusal to see more than is displayed.

While their UK albums tended to show the lighter side of Laibach's music, more atmospheric, and distinctly non-rock oriented efforts continued to appear on the continent - the German soundtrack Baptism, with its heavy-handed reworking of **Queen**'s "One Vision" is another example of the band's playfulness, while their two swipes at western culture, tackling the **Beatles**' Let It Be album in its near-entirety, and the **Rolling Stones**' "Sympathy for the Devil", are perhaps their most effective gestures yet - particularly the Stones track, repeated six times across the course of one album.

These releases marked the conclusion, at least temporarily, of Laibach's most obvious pretensions to pop music; their next album, in 1989, was the soundtrack to NSK's revolutionary production of Shakespeare's MacBeth, an album widely (but not necessarily accurately) described as their best. Kapital, however, reverted to recent type, a club scene suitor which only partially works.

Discography: Laibach/The Last Few Days (Skuc 1983); Boji EP (Laylah 1984); Laibach (Skuc 1985); Through The Occupied Netherlands (Staal 1984); Rekapitulacija 1980-84 (WUS 1984); Vstajenje v Berlinu (Skuc 1984); Neue Konservativ (no label 1985); Life In Hell (V2 1985); Ein Schauspieler (Staal 1985); Nova Akropola (Cherry Red 1986); Divergences/Diversions (Le Reseau 1986); Occupied Europe Tour 1983-85 (Side Effects 1986); Geburt Einer Nation (Mute 1987); Slovenska Akropola (Skuc 1987); Life is Life (Mute 1987); Klangniederschrift Einer Taufe/Baptism below Trigloav (Sub Rosa 1987); Sympathy For The Devil 3xEP/LP (Mute 1988); Let It Be (Mute 1988); MacBeth (Mute 1989): Kapital (Mute 1992).

LARD see Ministry

LEAD INTO GOLD see Ministry

LEATHER NUN

One of the original releases on Industrial Records, dating back to 1978 this Swedish quartet released a live cassette (Live At Scala) with contributions from Genesis P. Orridge and Monte Cazzazza. Monte also produced their debut EP "Slow Death" (also on Industrial Records).

DISCOGRAPHY: SLOW DEATH (INDUSTRIAL 1978)

LEÆTHER STRIP

Germany's **Claus Larsen** remains amongst the most challenging artists currently working in the field of industrial/techno dance music. Having debuted with a demo version of the Pleasure of Penetration album, Leaether Strip signed to the German Zoth Ommog label, and released both a rerecorded Pleasure, plus the "Japanese Bodies" 12". Technopolis II and the "Aspects of Aggression" 45 followed.

Leætherstrip

Although Leætherstrip continued to build a firm support base in the US, it was not until 1992 that a domestic licensing deal was arranged with Re-constriction/Cargo. The Penetrate the Satanic Citizen compilation excerpts four past albums, plus "Leætherstrip II" from the "Aspects" 12", plus a cut contributed to the Body Rapture 1 sampler.

Legendary Pink Dots

Material is a seven track collection of live and studio cuts, and includes "Steal", Leaetherstrip's unused contribution to the 1992 Cool World movie soundtrack. (See **Klute**).

Discography (release dates unavailable): Aspects Of Agression EP (Zoth Ommog); Japanese Bodies 12" (Zoth Ommog 1989); Object V EP (Zoth Ommog); The Pleasure of Penetration (Zoth Ommog); Technopolis 2 (Zoth Ommog); Science for the Satanic Citizen EP (Zoth Ommog); Body Rapture EP (Zoth Ommog): Technopolis 3 (Zoth Ommog); Solitary Confinement (Zoth Ommog); Penetrate the Satanic Citizen (Re-Constriction 1992); Material (Zoth Ommog 1992); Fit For Flogging (Cleopatra 1993).

LEAGUE OF GENTLEMEN see Robert Fripp, XTC

THOMAS LEER

Utilzing his arsenal of primitive electronics in pure DIY fashion, Thomas Leer first emerged on the wide screen as one of several avart-gardists on Cherry Red's seminal Business Unusual compilation in January 1979 - also included were **Cabaret Voltaire**, **Robert Rental**, and **Throbbing Gristle**. Leer's contribution was the insectoid "Private Plane", previously available (c/w "International") on a desperately obscure 1978 single.

The Bridge, Leer's groundbreaking collaboration with **Robert Rental**, followed; scarcely the most prolific artist on Cherry Red's books, Leer then resurfaced on another pair of compilations - "King of Sham" turns up on Perspectives and Distortions; another piece apparently appears on Morocci Klung!, with the 4 Movements EP proving that he could sustain his work over more than two songs at a time - the EP has four.

The double 12" single Contradictions completely shattered Leer's reputation as a minimalist, and the arrival of "All About You"/"Saving Grace" in November rounded off Leer's most prolific year yet. A 1992 CD reissue for The Bridge aside, he has remained silent ever since. (**see Robert Rental**)

Discography: The Bridge (with Robert Rental Industrial 1979); 4 Movements EP (Cherry Red 1981); Contradictions EP (Cherry Red 1992).

LEGENDARY PINK DOTS

"We tried to book a show in Austria and the guy in the club thought it was a hoax because I had committed suicide and the band had split up" - **Edward Ka-Spel**, 1991

Rooted firmly in the post-hippy aftermath best associated with **Julian Cope**, **Robyn Hitchcock** and occasionally **Robert Smith**, London's Pink Dots - fronted by the self-styled enigma **Edward Ka-Spel**, and **Phil Knights** - employ sampling, found sounds and a bizarre imagination to conjure up a modern approximation of **Magma**.

Homespun legends rooted in destruction and forboding have fired the Pink Dots repertoire since the band's inception in 1981, with the self-released Chemical Playschool and Kleine

Kreig cassettes. Since then, well over a dozen albums (and as many members) have traced the duo's convoluted imagery, most of which are thoroughly representative, in the eyes of their fans if not the US immigration authorities. American tours in 1989 and 1991 sandwiched an aborted venture in 1990 - the band was turned down for visas on that occasion, due to "lack of artistic merit." "I can just see them all in the immigration department bopping away, going 'ah, the Pink Dots have redeemed themselves'," Ka-Spel later remarked.

The anthology Stone Circles, and the triple Legendary Pink Box (comprising out-takes, alternate versions and rarities) make a handy point of entry for the uninitiated; of recent releases, the haunting Shadow Weaver is the first part of a concept album; Malacahi, unreleased at the time of writing, completes the offering.

Ka-Spel has also released one (irrelevant) solo album, and performed with **Celvin Key (of Skinny Puppy)** on the Teargarden release Tired Eyes Slowly Burning (Nettwerk 1987). (see **Skinny Puppy**).

Selected discography: Chemical Playschool (Microdot 1981); Kleine Kreig (Microdot 1981); Brighter Now (Terminal Kaleidoscope 1982); Curse (Terminal Kaleidoscope 1983); Faces in the Fire (Terminal Kaleidoscope 1984); The Tower (Terminal Kaleidoscope 1984); Greeting 9 (Materiali Sonori 1984); The Lovers (Ding Dong 1985); Asylum (Play It Again Sam 1985); Island of Jewels (Play It Again Sam 1987); Stone Circles (Play It Again Sam 1987); Any Day Now (Play It Again Sam 1987); The Golden Age (Play It Again Sam 1988); Legendary Pink Box (Play It Again Sam 1989); Crushed Velvet Apocalypse (Play It Again Sam 1990); The Maria Dimension (Play It Again Sam 1991); Shadow Weaver (Play It Again Sam 1992); Malacahi (Play It Again Sam, 1993).

LORDS OF ACID

Hard Techno-Industrial crossover, minus the disposableness. Appeals to a lot of Industrial fans who dont like Techno generally.

Lords Of Acid

LOOP see **Godflesh**

LOVE STREET see **Stephen Mallinder, Soft Cell**

LUNG

Vancouver based noisemakers whose first two singles, 1991's "Psychopornodelia" and 1992's "Litany" (a UK Alternative chart hit) set the stage for a devastating 1993 debut, Magnum Opiate, co-produced by **Don Gordon** of **Numb**.

Among the bands other achievements can be counted a statewide ban in Washington. (See **Numb**).

LYDIA LUNCH see **Einsturzende Neubauten, Foetus, My Life With The Thrill Kill Kult**

MAGMA

Although the joke swiftly (three albums) wore thin, Magma were at one point perhaps the grandest and most exquisitely executed hoax of the entire pre-industrial era.

Most tellingly described as the French **Hawkwind**, the group - under the dictatorial leadership of **Christian Vander** - was formed to tell, or perhaps warn, of earth's demise, the result of a lengthy dispute with the otherwise unknown planet Kobaia. Smacking more than a little of an id image of **Gong**'s hippy-trippy pot-head pixie kick, Magma could nevertheless prove remarkably inspired; lyrics were sung in the band's own made-up language, and the music adhered to similarly personal, and generally unrecognizable, reference points.

A 1973 press hand-out, accompanying Mekanik Destructive Commandoh, claimed for Magma "a regular maelstrom of influence", namedropping **Bartok, Stockhausen, Soft Machine** and Hair, a somewhat scarifying combination which "demands of the listener as much as it gives him, more than just a physical presence and some vague wailings. Here, opening the door to anguish, many questions are asked... expressing the inescapable refusal, like violent vomiting, of a world where nothing is worth living anymore because there isn't a single spark of purity left." Sounds familiar?

Selected discography: Magma (Phillips 1970); 1001 Degrees Centigrade (Phillips 1971); Mekanik Destructive Kommandoh (A&M 1973); Wuroah Itah (Barclay 1974); Köhntarkosz (A&M 1974); Live (Utopia 1975); To Love (Seventh 1991).

STEPHEN MALLINDER

The **Cabaret Voltaire** founder released his first solo record in 1982. Both the LP Pow Wow and the 12" single "Temperature Drop", were major indy chart hits, and three years later were packaged together as Pow Wow Plus.

While partner **Richard Kirk** remained based in the Cabs' hometown of Sheffield, Mallinder moved to London and in 1988, he launched a new band, **Love Street**. Also involved were **Soft Cell** keyboard wizard **Dave Ball** and **Mark Brydon**. Love Street released just one single, "Galaxy"/"Come On Down To Love Street". Mallinder also figures in the **Ministry** side project **Acid Horse**. **(see Cabaret Voltaire, Hafler Trio, Richard Kirk, Ministry, Soft Cell)**

Discography: Pow Wow (Fetish 1982); Pow Wow Plus (Doublevision 1985).

MARIAS BAD

A leading light in the Norwegian industrial scene, Marias Bad debuted in 1991 with the one track cassette Oneidro Noaida, a drums and bass oriented piece over which are layered choirs and a whale.

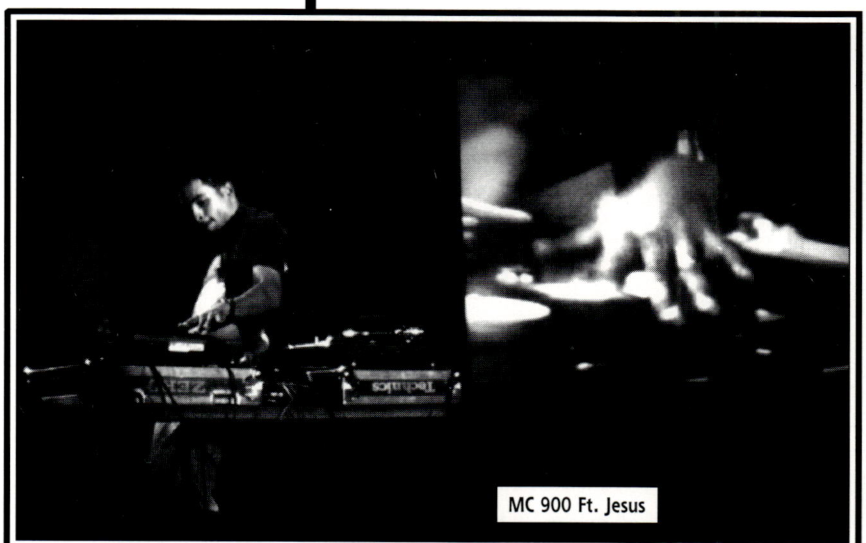

MC 900 Ft. Jesus

MC 900 FOOT JESUS

"I get bored really fast" - **Mark Griffin**

Dallas born Griffin is the joker in the industrial pack, utilizing samples and sounds, not to mention a band name which are guaranteed to provoke some kind of reaction. Musically limited (the debut EP obsesses on **Public Enemy** to the point of distraction), Jesus remains both challenging and fun, lacking only the focus to sustain his strengths over an entire album.

Discography: With DJ Zero EP (Nettwerk 1989); Hell With The Lid Off (Netwerk 1990); Welcome To My Dream. (Nettwerk 1991).

MEAT BEAT MANIFESTO

MBF launched themselves out of London's terminally turbulent acid rave scene in 1988 with a string of powerful dance singles, each backed up with a similarly inventive series of live and video performances; emphasizing costume and choreography as an integral part of the band, with a success unseen since the heyday of British Glam (an undeniable influence on the group), the band was about to release its debut album when the tapes were lost in a fire.

Instead, four past singles were substantially remixed to create the double Storm The Studio album; material surviving from various stages of the doomed first album was meanwhile cleaned up and released, with suitable warnings, as Armed Audio Warfare, an album whose unfinished nature in no way detracts from its power.

Meat Beat Manifesto

Meat Beat Manifesto

Concentrating as much, if not more, on sampled and found sounds as they do on their own musical contributions, MBF neverthless rise above accusations of simple plagiarism by the sheer inventiveness of their work - like the **Justified Ancients of Mu**, Meat Beat Manifesto create dynamic sound sculptures, with 99% - the only MBF album to have been recorded and released according to plan! - a stark vindication of their chosen method of working.

Selected discography: Storm The Studio (Wax Trax 1989): Armed Audio Warfare (Wax Trax 1990): 99% (Elektra 1990): Now EP (1991)

MENTALLO & THE FIXER

Texans **Garry Dassing** and his brother **Dwayne** formed the bizarre Mentallo in 1990 following vocalist **Richard Mendez**'s departure from an earlier project, **Benestrophe.** Almost completely computer-based, they debuted with a cassette shared with **Non-Aggression Pact** (General Purpose Cassettes), received with unanimous joy throughout the industrial fanzine scene.

Released in Europe only, by the Portuguese Simbiose label, the debut album, No Rest For The Wicked was compiled from three years worth of material. In 1993 Zoth Ommog released Revelations 23.

METAL MACHINE MUSIC

It's 1976 and **Lou Reed** is a star. **The Velvet Underground**, the New York band he powered through four years of obscurity, is a legend; "Walk on the Wild Side", his solo hit from three years back, is an AM staple and will one day sell Suzuki motorcycles. His last album made the Top 10 and now his record label is on the phone, demanding a follow-up.

"Okay," says Lou. "I can do that."

He rummages around for a cassette player. "I've got something here I wanted to do way back when. I had to wait a while to get the equipment, but now I've got it and it's done." He depresses the Play button. "It's a double album. Wanna hear it?"

The record company man can scarcely control his excitement. "Are you kidding? Hey, c'mon Lou baby, let me have it."

At the other end of the line, Lou smiles. "Okay."

For a moment, silence. Then it begins. ZZZZZZZRRRRRRRRRRRRREEEE EEEEEEEGGGGGGGRRRRRRAAAAAAAAARRRRRRRRRZZZZBBBBBBBBBB. By the time Lou's ready for side two, the line's dead. By the side he reaches three, so is his commercial potential.

A lot of lines went dead when Metal Machine Music was released. A lot of loyalties, too. Four sides of eardrum shattering feedback, Lou could blather forever about it being the ultimate guitar solo... tell folk that if they listened hard enough they'd hear all sorts of little tunes dancing around in there... he could even say they should be grateful he released it in the first place.

But when you came down to it, Metal Machine Music hurt. And not in a wimpy "Oh my God, that's so awful it's painful" way, it hurt in a "Oh shit, I think my brain just exploded" way. Most people returned their copy before they got to the end of side one. Others got to side four, then returned it.

Lou Reed was the first major artist to truly bring electronics into play, maintaining **LaMonte Young**'s pioneering spirit of atonal torment which had first inspired the Velvets (and which feeds throughout their most important, first two albums), a full two years before **David Bowie** (who produced "Walk on the Wild Side") and **Brian Eno** gave vent to their own interests. Even today, Metal Machine remains a massively influential album; even today, much of what Reed created on it has yet to be assimilated. **(see John Cale, Velvet Underground, LaMonte Young)**

Discography: Metal Machine Music (RCA Red Seal 1975); Metal Machine Music (Great Expectations 1990)

DANIEL MILLER

"I can never take anything I do totally seriously" - Daniel Miller 1978

It was the lowering price of synthesizers, and their increased visibility in the rock scene, which first attracted Daniel Miller to the instrument. As guitarist in a school band, he had continually annoyed his fellow musicians with his insistence on trying to wring fresh sounds from his instrument, usually by hitting it with things; later, as he watched the burgeoning punk scene in England, he prophesized that the synth would eventually replace the guitar as the ultimate punk instrument. Not only did you not have to be able to play guitar, he argued, with a synth you didn't even need to know how to hold one.

He proved his point in 1978 when he recorded and released "TV OD"/"Warm Leatherette", under the name **The Normal**. The debut release on his own Mute label, the single sold over 40,000 copies, persuading Miller that there was indeed a market for a label which released only experimental electronic sounds.

The Normal followed this project with the similarly harsh Live EP, recorded at Rutland Pavilion in March 1979 with fellow electronics buff **Robert Rental**. Then, while **Fad Gadget** and **Boyd Rice** kept the early Mute catalog moving, Miller moved on to **the Silicone Teens**, a largely successful attempt to crush the essence of rock'n'roll into a tight electronic corner, in which even the harshest lyric emerged as a squeaky synthesized gurgle.

Miller's discovery of the nascent **Depeche Mode**, and their ensuing explosion into popularity appears to have stripped much of the urgency from Miller's own work; he has not recorded in his own right since the Teens' album, although he has surfaced for occasional one off projects - the gripping "Or So It Seems"/"Heart of Hearts" single, recorded under the alias of **Duet Emmo** with **Dome**, and remix work for **Chris and Cosey** for instance.

Nevertheless, he remains a vibrant, and vital, figure on the electronic industrial scene. (**see Chris & Cosey, Depeche Mode, Dome, Punk Rock, Robert Rental, Boyd Rice, Frank Tovey**)

Discography: Live EP (with **Robert Rental**–Rough Trade 1980);
SILICON TEENS: Music for Parties (Mute 1980).

MINISTRY

by Jo-Ann Greene (**originally published in Goldmine magazine**)

In the beginning there was Cuban-born **Al Jourgensen**, who at age one fled the incoming Castro regime with his family. His parents divorced soon after, and it is his step-father's name that Jourgensen has retained. Al spent his teen years living in Denver, and claims to have taught himself bass by listening to **Syd Barrett** and **Hawkwind**. The influence of the latter, with the driving roar of **Lemmy** firmly to the fore, has remained with Jourgensen ever since.

Ministry (Marsden Daley, Paul Barker, Bill Rieflin, Al Jourgensen)

More diverse influences came from the power of **Led Zeppelin**, and the angst of country music, but it was the **Ramones** who directly changed Jourgensen's life. Seeing them live convinced Jourgensen to form his own band, and he fully

Special Affect (Henry Rushakoff, Al Jourgensen, Frankie Nardiello, Marty Sorensen)

embraced punkdom in all its glory.

Moving to Chicago, Jourgensen went through a number of teenaged bands before forming **Special Affect** around 1978, with vocalist **Tom Hoffman**, bassist **Marty Sorensen**, and drummer **Harry Rushakoff**. (Hoffman was later replaced by **Frankie Nardiello**.) Two singles, an EP and a soundtrack album followed, before Special Affect folded in 1981: drummer Rushakoff moved on to **Dream Six** (the embryonic **Concrete Blonde**); vocalist Nardiello later resurfaced in **My Life With The Thrill Kill Kult**. Jourgensen renamed himself Ministry and landed a deal with the nascent Wax Trax label.

Operating out of an office on Lincoln Avenue, Chicago, Wax Trax -like Rough Trade and Chiswick in Britain several years earlier -grew out of a record store, owned by **Jim Nash** and **Dannie Flesher**.

Sensing the popularity of British technopop both on the import racks and the dance floors, Nash and Flesher dedicated their label to homegrown answers to the new musical rage, and Ministry swiftly repaid their faith.

Ministry - Jourgensen, plus bassist **Lamont Welton**, a horn player named **Preston**, **Steve Brighton** on backing vocals and drums by **Stevo**, Britain's Some Bizzare label's enigmatic supremo - debuted late in 1981 with a 12" single, "I'm Falling", backed by both a regular and a dub mix of "Cold Life", and "Primental".

Combining club-pop sensibilities with what in later years would be described as a light industrial feel, but which in 1981, was simply harsh techno, "I'm Falling" was an immediate hit on the club circuit, even securing a U.K. release when Wax Trax leased it to Situation Two. By mid-1982, Ministry - which now comprised Jourgensen and drummer **Stephen George** - had signed to Arista.

Jumping on the New Romantic/synthi-pop bandwagon, Arista had high hopes that in Jourgensen, they had discovered the American **Howard Jones**. Young and impressionable, Jourgensen allowed himself to be pushed by the company executives into making an album he now despises. It was a lesson he was never to forget.

Ministry

With Sympathy appeared in mid-1983. Co-produced by **Psychedelic Fur Vince Ely**, the album has been Jourgensen's cross to bear ever since. Lusher and lighter than "I'm Falling", the album and its attendant singles were a smash on the dance floor, breezy electronic pop which was perfectly in keeping with the fashions of the day, but lost its appeal as swiftly as those fashions. Today, Jourgensen refers to it as a "complete abortion and total piece of corporate shit", proudly adding that "the difference between Ministry and other bands is that we sold out before we even started."

At the time, however, Jourgensen played the corporate game to the hilt. Ministry toured with **Culture Club** and **Depeche Mode**, and when he arrived in London that fall, he complained not about his loss of artistic freedom, but about the unfair treatment Ministry

Alain Jourgensen

received from the US press, simply because it was American.

Nevertheless, Jourgensen's relations with Arista became increasingly acrimonious as time passed, until reaching a head when Al submitted a demo version of a ridiculously feather light cover of **Roxy Music**'s "Same Old Scene". Arista dropped Ministry soon after.

Returning to Wax Trax, Jourgensen determined to remake his career in his own chosen image. It was to be an uphill battle, complicated by his decision to keep the Ministry moniker, but by early 1985 the attack was underway.

A reissue of the "Cold Life" 12" served notice of Ministry's return, followed swiftly by a new 12", "All Day"/"Everyday is Halloween". To help exorcise the ghost of the intervening years, Jourgensen even called back Stevo, his co-conspirator for "Cold Life", as guest percussionist.

Alain Jourgensen

A third single, "The Nature of Love", appeared before the end of the year.

Jourgensen was taking a leap into the musical darkness. Producing, writing, and creating the majority of the music himself (Stevo aside, the only other "musicians" in sight were wife **Patti**, and **Richard 23**, a member of Wax Trax labelmates, Belgian synth rockers **Front 242**), the singles were an angry statement of intent. In "Everyday is Halloween", for instance, he demands "Why can't I live a life for me, why should I take the abuse that's served?" It was a far cry from "Work for Love"

Although much of its impact has been deadened by time, Twitch remains one of the most uncompromising albums of the 1980s. It was produced by **Adrian Sherwood**, the London based producer best known for the dense, dub-laden "On-U" sound which he named after his own record label. But he would do for Ministry what he done for Depeche Mode the year before. Previous to their work with Sherwood, Depeche were regarded as basically a teeny-bop synthi-pop band, a tag which they resented almost as much as Jourgensen hated his own reputation.

A tour was planned, but what Jourgensen could create in the studio with the aid of a few additional musicians, he could not recreate live without a band. This gap was filled by the recruitment of Roland and Paul Barker, and Bill Rieflin, members of the recently deceased **Blackouts**, a Seattle No Wave band whose final single, "Lost Soul's Club", Jourgensen produced for release on Wax Trax. Paul Barker, also a former member of Seattle synth pioneers **Young Scientists**, and Rieflin, have remained with Jourgensen ever since. (The lighting man on the tour, incidentally, was Frank Nardiello, Special Affects' vocalist.)

Jourgensen and Paul Barker, meanwhile, headed back to the Wax Trax studio, where they set about recording the first of the many Ministry side projects. Named for a street gang in Brussels, the **Revolting Cocks**' floating line-up has, at varying times, included the likes of **Chris Connelly**, formerly of the Scottish band **Fini Tribe**; **Luc van Acker**, the Belgian musician/producer who was one of several guest musicians on Twitch, Richard 23, and **Jeff Ward**, who would later join Nardiello in **Thrill Kill Kult**.

Rev Co (to which the band's somewhat controversial name is frequently abbreviated) released two singles in 1986, "No Devotion" and "You Often Forget", the latter in two self-explanatory mixes - a four minute "Benign" version, and a ferocious eight minute "Malignant". A Cocks album, Big Sexyland, also appeared. As for the band, Jourgensen has referred to them as a "lampshade and limbo-line party - put on a lampshade, do the limbo, go to a party and out comes a record...."

Alain Jourgensen

Then it was time to get serious, Ministry were about to embark on the recording of their new album, when Sire contacted Jourgensen. They had a few suggestions they wanted to make, in the hope that Al could perhaps make a record somewhat less violent (and more commercial!) than

Twitch. Jourgensen's reaction was short and sweet - "No!"

In retrospect, although Twitch was an improvement on his past records, Jourgensen was still not entirely happy. Now, however, he was finally in the perfect situation, with no-one to answer to but himself. For the new album, Ministry wouldn't even bring in an outside producer; Jourgensen and Paul Barker (under the aliases **Hypo Luxa** and **Hermes Pan**) would do it themselves.

It would take two years for The Land of Rape and Honey to be completed and see the light of day, a lay off which was broken only by the release, on Wax Trax, of the Twelve Inch Singles 1981-84 album in 1987. A 55 minute CD compilation which brought together all four of Ministry's pre-Twitch single A-sides, plus two b-sides - "Cold Life Dub" and "Everyday Is Halloween", and fresh remixes of the "Nature of Love" Cruelty Mix and "All Day Remix", from Twitch itself.

The Land of Rape and Honey finally appeared in October, 1988, accompanied by a 12" single offering a remix of "Stigmata" and "Tonight We Murder". In its wake, three other spin-offs were formed and released records that year.

The Cocks themselves released a new single, the three track "Stainless Steel Providers". There was also a double live album, and a video, recorded at a riotous show in Chicago in September, 1987, and both titled You Goddamned Son of a Bitch.

Then came **10,000 Homo DJs**, who released their first single, "Better Ways". The DJs were actually Jourgensen, Barker, Rieflin, and **Mike Scaccia** (who would himself become a semi-permanent member of Ministry), and the single was an outake from the Rape and Honey sessions which Jourgensen felt was inappropriate for release under the Ministry moniker.

Revolting Cocks (William Rieflin, Chris Connelly, Luc Van Acker [front], Al Jourgensen, Paul Barker

Pailhead - Jourgensen, Barker and **Fugazi**'s **Ian MacKaye** - was next. A single, "I Will Refuse" was followed by the Trait EP, released on cassette and CD with both sides of the "I Will Refuse" 45 as bonus cuts).

Yet even all these projects were not sufficient to sate Barker's creative thirst. Under the pseudonym **Lead into Gold**, he released his debut solo single, a rerecording of the Blackouts' b-side "The Idiot".

Neither was this explosion of activity confined to the Wax Trax studios, as Jourgensen's guest appearance on **Skinny Puppy's** Rabies confirms. (**Ogre** repaid the favor by appearing with Ministry on stage; additionally, **David Ogilvie**, Skinny Puppy's long-time producer, has also worked with both Ministry and the Revolting Cocks.)

With the release of Rape and Honey, Ministry went back out on the road, accompanied by a large coterie of musician friends - a veritable who's who of the alternative scene: Chris Connelly and Ogre were joined by **Nine Inch Nail**'s **Trent Reznor**, **Jesus Lizard**'s **David Yow**, **William Tucker** (ex-**Scornflakes**), **Lab Report**'s **Matt Schultz**, ex-**Discharge/UK Subs Terry "Tezz" Roberts**, and ex-**PIL/Killing Joke** drummer **Martin Atkins**. Opening for the Ministry circus was German industrialists **KMFDM**, and they, too, joined the headliners on stage on occasion.

Paul Barker

1989 brought further chaos. Ministry released a new album, The Mind is a Terrible Thing to Taste, with Chris Connelly again taking the mike, handling lead vocals on two tracks, "Cannibal Song" and "Never Believe", and another tour, again with a massive supporting cast. Highlights from this outing are included on the six track/40 minute In Case You Didn't Feel Like Showing Up live album (the video of the same name features two extra tracks), released the following year.

There was also a new 12" single from the Revolting Cocks, featuring two different mixes of "Physical" - "(Let's Get"), and a monstrous 10 minute "(Let's Talk)"; then, utilizing musicians who appeared on the tour, three further spin-offs formed in 1989: **Acid Horse**, **PTP** and **Lard**.

Acid Horse is the intriguing combination of Barker/Jourgensen and **Cabaret Voltaire**'s **Richard Kirk** and **Stephen Mallinder**, going under the bizarre monikers of Harold Sando, Alien Dog Star, Tennessee King, and Biff "Gallopin" Scorpio Saddlebutt. "No Name No Slogan" was produced by Jourgensen/Barker once again under their Hypo Lux and Hermes Pan aliases. The remix on the B-Side is by Cabaret Voltaire.

The song is strangely reminiscent of **Blancmange**'s "Blind Vision", if it had been remixed for inclusion on the soundtrack of The Good, the Bad and the Ugly. For the B-side, imagine the same performed by **Kraftwerk**.

PTP has been described by Jourgensen as "a total lark, our chance to go in and take a piss out of the old house-music joke. How many people does it take to make a house record? Seventeen. Sixteen to go on a liquor run and one to push the button." The band itself comprises Scorpio, Alien Dog Star and Frenchie l'Amour - a.k.a. Barker, Jourgensen, and Connelly, while the solitary PTP single "Rubber Glove Seduction" is an hilarious song involving alarm clocks, a synchopated drum beat and Connelly in monotone repeating "Tick tick tock, I am the kitchen clock...."

The b-side, "My Favorite Things" moves from House to Rave, in which Connelly in a breathy whisper imparts "My favorite things." Was Hollywood in on the joke when they asked PTP to contribute a song for the Robocop soundtrack?

Finally there was Lard, starring Alien Jourgensen, Barker, Rev Co.'s Jeff Ward, and the **Dead Kennedy**'s **Jello Biafra** - who can be heard on In Case..., reciting the flag pledge. Time to Melt, Lard's debut EP, was also released that year, and it's in Lard that Jourgensen's and Barker's thrash material have found a home.

At Wax Trax the party never seemed to end. Jourgensen has often been quoted as preferring studio work to the rigors of the road. And why not, when the studio is a non-stop musical celebration? Friends are always around, people are always dropping in, and the musical juices never stop flowing. It was this spirit that was to see the creation not just of the Ministry albums, but of the spin-off bands as well.

"People used to stop by the studio late and night," Jourgensen explains, "and we'd come up with real crazy stuff. We'd have painters and poets and musicians coming over and we'd just record the party." The problem was, not all of it was appropriate for Ministry; thus the formation of the spin-offs.

1990, in comparison to the two previous years, was a thin year for Ministry related material.

The live album album aside, there were no other Ministry releases at all, although the 10,000 Homo DJs did resurface with a jarring cover of **Black Sabbath**'s "Supernaut" like "Better Ways" an outtake from the most recent Ministry album, and again released in 12" form only. The DJs' absence from the CD/cassette scene was eventually remedied with the compilation of both sides of both 12 inchers onto the Supernaut EP.

Revolting Cocks (Richard 23, Luc Van Acker, Al Jourgensen)

There was also a new Revolting Cocks album, the provocatively titled Beers Steers and Queers, as well as a twin remix single of the title track (to which was added live renditions of the 1988 Revco single "Stainless Steel Providers", and **John Lydon**'s "Public Image" anthem).

Finally, Barker's Lead Into Gold released the Chicks and Speed = Futurism EP as a five track 12", and - with the addition of "The Idiot" - on cassette and CD. The long awaited Lead Into Gold album, Age of Reason, followed.

1991, too, passed with little apparent activity, with only Lard (whose album The Last Temptation of Reid includes a riotous cover of **Napoleon XIV**'s "They're Coming To Take Me Away Ha Ha") breaking the silence. And then, in October, Ministry released "Jesus Built My Hot Rod", the single which at last would gain them serious recognition, and a song that epitomizes the Jourgensen approach to song creation. "Originally, I was going to sing that song", Jourgensen said in an interview, "and I had the lyrics for it, but then **Gibby** (**Haynes**)

Al Jourgensen

David Oglivie

Alain Jourgensen

came to the studio (with the rest of the **Butthole Surfers** who were in town with the first Lollapalooza tour) got tanked up on tequila, sang his vocal over the bass line and it just kind of digressed from there."

Jourgensen has joked that they were going to include a lyric sheet with the record, compiled phonetically, as even he can't make out what the words are. "Gibby really just invented his own language for the song, that's the only way to describe it. "

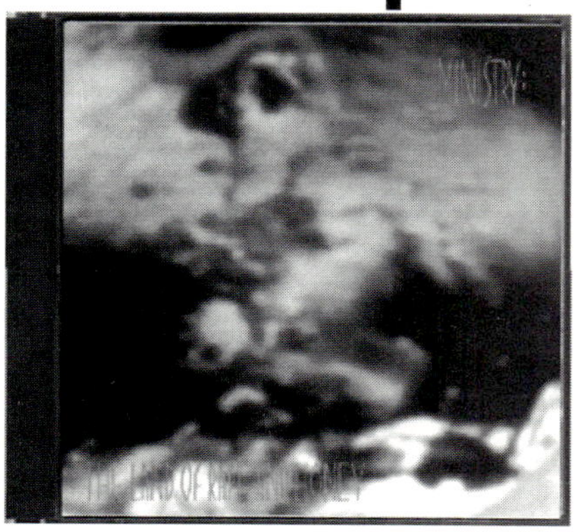

An immediate college chart hit, "Hot Rod" ensured Ministry a place on the next Lollapalooza tour, despite Jourgensen's reservations. Desperate to hit on a good excuse not to play, he admits that he made as many unreasonable demands as he could think of, in hopes of the tour falling through. To his surprise, the organizers agreed to every one of them, including Al's demand that Ministry take the stage after sunset - ensuring they were given the second headliner slot. Ministry had arrived.

Back at Wax Trax, however, the musical free-for-all of the past was turning into an after-hours club. Jourgensen and Barker were running out of patience fast. "These days people come over for that last beer of the night. There's no creativity, just a bunch of people looking for the party." The band started locking the studio doors at 4 a.m. - as the clubs closed - to keep out the riffraff.

Finally, Jourgensen and Barker had enough, and with Bill Rieflin in tow (Mike Scaccia was amongst the other auxiliary Ministers on the completed album), moved north to Royal Recorders' studio at Lake Geneva, Wisconsin. For ten weeks the Ministry core worked and worked, but to little avail. The enthusiasm began to disipate quickly. Eventually they returned to Chicago to finish the album. In all, Psalm 69 - The Way to Succeed and the Way to Suck Eggs would take a grand total of fifteen months.

But as usual, there was more than a Ministry album to be found amongst the some thirty songs that were eventually taped. Nine ended up on Psalm 69 itself; the remainder are to be distributed amongst the myriad of side projects. One, a fresh version of "Supernaut" with guest vocals by Trent Reznor, has already leaked out, courtesy of a joint Wax Trax/Play It Again Sam sampler.

That was the good news, there was some bad news as well. The album came in three times over budget (a whopping $329,000 - alright for **Fleetwood Mac**, but far from acceptable for a still relatively unknown "alternative" band.) Then there was the surprise of discovering that the new album was virtually a heavy metal offering - although heavy metal is hardly accurate: wall of noise might be more appropriate.

But if Ministry had doubts, the public didn't. Psalm 69 entered the Billboard chart at #27 and the CMJ chart at #1. The initial shipment of 250,000 copies was almost 100,000 more than any of the previous albums. And if Sire were uncomfortable with the band's political intent, they needn't have worried. The public agreed with them entirely. "N.W.O." became one of the most requested alternative hits of the summer of 1992, without even being released as a conventional single, while the anti-drug "Just One Fix" followed it into heavy rotation on radio and MTV alike.

Ministry are currently in the process of relocating to Austin, Texas, home of the Butthole Surfers, where they intend to set up their own studio and record label. (see **My Life With The Thrill Kill Kult, Nine Inch Nails, Pigface, Skinny Puppy, Synthipop**)

Select discography: MINISTRY With Sympathy (Arista 1983); Work For Love (Arista UK 1983); Twelve Inch Singles 1981-84 (Wax Trax 1985); Twitch (Sire 1986); The Land of Rape and Honey (Sire 1988); The Mind Is A Terrible Thing To Taste (Sire 1989); In Case You Didn't Feel Like Turning Up (Sire 1990); Psalm 69 (Sire 1992)
ACID HORSE: No Name No Slogan EP (Wax Trax 1989)
LARD: Power Of Lard EP (Alternative Tentacles 1989); Last Temptation of Reid (Alternative Tentacles 1991)
LEAD INTO GOLD: The Idiot EP (Wax Trax 1988); Low and Slow EP (Wax Trax 1988); Chicks and Speed - Futurism EP (Wax Trax 1990);
Age Of Reason (Wax Trax 1990)
ONE THOUSAND HOMO DJs: Supernaut EP (Wax Trax 1992)
PAILHEAD: Trait EP (Wax Trax 1988)
PTP: Rubber Glove Seduction EP (Wax Trax 1989)
REVOLTING COCKS: No Devotion EP (Wax Trax 1986); You Often Forget EP (Wax Trax 1986); Big Sexyland (Wax Trax 1986); Stainless Steel Providers EP (Wax Trax 1988); You Goddamned Son of a Bitch (Wax Trax 1988); Physical EP (Wax Trax 1989); Beers Steers and Queers (Wax Trax 1990); Beer Steers and Queers EP (Wax Trax 1990)

MOEV

Contemporaries of the better-known **Images in Vogue**, this pioneering Vancouver-based electronic band released a number of independent records in Vancouver, before moving to San Francisco and a shortlived deal with Go Records. They then returned to Vancouver, where guitarist **Ric Arboit** and manager **Terry McBride** co-founded of Nettwerk Records, for whom Moev continued recording.

Selected discography: Cracked Mirror EP (indy 1981); Zimmerkampf (Go Records 1982); Rotting Geraniums (Go Records 1983).

GIORGIO MORODER see Human League, Kraftwerk, Kraut Rock, Donna Summer

MURDER INC. see Pigface

MURDER IN EXILE

Formed in '91 by **Scott Franklin** and **Tom Leinbach** in New York City. Their project moved to L.A. the following year and hooked up with bassist **Zack Bezner**, sample keyboard wizard **"Otis"** and drummer **Doug Duffy** to complete the line-up, resulting in a wall of sound sure to piss off every American parent.

Select discography: Murder In Exile (Ratchet Blade 1991).

MY LIFE WITH THE THRILL KILL KULT

My Life With The Thrill Kill Kult (Frankie Nardiello, Marsden Daley)

An irreverent cross between **Tubes** and a funky **Ministry**, fired by a theatricality which put even **Funkadelic** in the shade, TKK hit paydirt when "Sex On Wheels" was included on the soundtrack of the Cool World movie - a wild video featuring scantily-clad women doing exotic things in sunglasses, tales of a Sexplosion dwarf revue and (of course) a stunning song combined to turn a well-kept secret into a national phenomenon. And all without a single mention of Industrial Vaudeville.

The band came about when **Frank Nardiello (Groovie Mann)**, Ministry's lighting man on an early 1980s tour, and vocalist with **Al Jourgensen**'s pre-Ministry band **Special Affect**, met with Boston punk veteran **Marston Daley (Buzz McCoy)**. Working originally in video, they fell into music only when they needed to compose a soundtrack for one such; encountering Chicago's Wax Trax label, they landed a one-off EP deal, "and it just took off from there." According to Daley, "we just never got round to doing our films."

My Life With The Thrill Kill Kult

A club hit "Cuz It's Hot", recorded with **Lydia Lunch**, and a 1990 12" split with **KMFDM** provided early introductions to a band whose uncompromising slant on electronic funk had also been spread across a brace of albums: tours with **KMFDM** and **Pigface** (ex-TKK sideman **William Russell** also appeared with the latter added to the band's appeal, so that by the time of Sexplosion's appearance, the scene was set.

Comparable in many ways to the **Red Hot Chili Peppers**, TKK have nevertheless forged a unique direction for themselves; barring a descent into self-parody or more, it should remain a fascinating diversion for many years more. (**see KMFDM, Ministry, Pigface, Psychic TV, Sleep Chamber**)

Selected discography: Sexplosion (Wax Trax 1992)

MY SIN

The music of My Sin exists on the outer fringes of what can rightly be termed "industrial." With a jerry-rigged bank of electronics doing an uncanny impersonation of a punk band behind him, My Sin seems more concerned with utilizing technology for artistic self-containment than for setting new trends in modern music.

My Sin

My Sin first emerged in 1983 from out of San Francisco's diverse underground scene. Conspicuous for always performing alone, he attracted something of a cult following on the West Coast about the time of his first release, the somewhat disappointing Beyond Good And Evil EP. My Sin shows were notable for their usage of video to underscore his pounding, distorted synth-driven songs. In 1987 he released the striking They Rode Me Out Of Town On A Rail, a furious, chaotic foray into the world of techno-thrash. Less forceful, but still containing many engrossing moments, is 1991's Tribes CD.

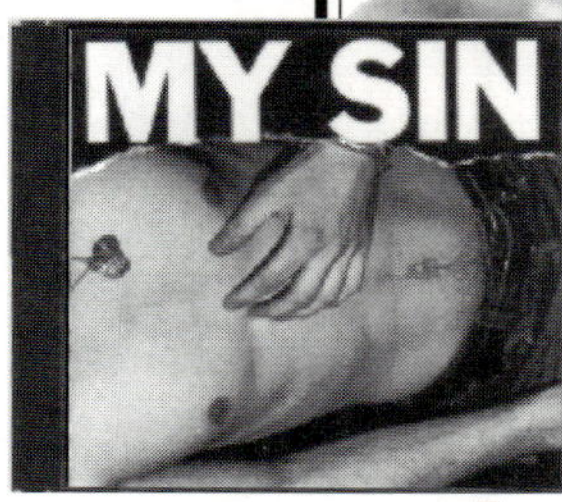

From the stripped down **Suicide**-ish minimalist pop attack of his early sound, to the neo-commercial fluff of Beyond Good And Evil, to the hard edges of They Rode Me Out Of Town On A Rail and beyond, My Sin's 10 year history has seen many phases, almost to the point of losing focus. His musical output is always stubbornly unique, probably owing to his rigid policy of never collaborating with other artists – even to the point of insisting upon engineering all his own recordings.

Discography: Beyond Good And Evil (Endless 1985); They Rode Me Out Of Town On A Rail (Endless 1987); All The Pussy You Want In A Dodge (Endless 1988); How To Talk Dirty And Alienate People (Endless 1989); Build My Gallows High (Endless 1990); Tribes (Endless 1991); Hot Rods (Endless 1993).

NEON JUDGEMENT

Early to mid-eighties Belgian Industrial-Goth cross-over duo (T.B. Frank and Dirk Da Davo).Brilliance in all if you avoid post-1989 recordings.

Selected Discography: The First Judgements (Pias); MBIH (Pias); General Pain and Major Disease (Pias); Horny As Hell (Pias).

Neu

NEU

Having been present at the birth and vinyl baptism of **Kraftwerk**, **Klaus Dinger** and **Thomas Homann** quit shortly after the release of that band's debut to form **Neu**, a band cut very much in the mold of early **Tangerine Dream**.

Pioneering both random electronics and ncise, the group's self-titled first album was produced by **Connie Plank**, and in the flurry of interest in the so-called Kraut Rock movement, was snatched up by UA, home a ready to **Can** and **Amon Duul II**. Another labelmate, **Dave Brock** of **Hawkwind**, penned liner notes to the album.

Neu 2 offered little variation on the themes espoused by its predecessor; however, the band's stark approach to both composition and recording continued to blaze trails some way above ones into which other bands were falling. **Neu 75**, a compilation of unreleased material recorded with additional members **Hans Lampe** (drums) and Klaus Dinger's brother **Thomas**, indicates that whatever prompted the band's demise, it was not a lack of ideas.

Prompted by the industrial/synthipop explosion of the early 1980s, the British Cherry Red label released Black Forest Gateau, a compilation of six tracks including one, "Hallogalld", which gave its name to a Germany-only album released in the mid-late 1970s. Two Originals repackages the first two albums. **(see Kraftwerk, Kraut Rock)**

Discography: Neu (UA 1972): Neu 2 (UA 1973); Neu 75 (Brain 1977); Hallogald (Brain 197?); Two Originals of Neu (Brain 1978); Black Forest Gateau (Cherry Red 1982)

NEW ORDER see Joy Division

NEWS FROM BABEL see Fred Frith

COLIN NEWMAN

Following the break-up of **Wire**, Newman launched into a short but generally interesting solo career, beginning with A-Z in October 1980. Comprised largely of material intended for the abortive fourth Wire album, the productive sessions spawned two excellent singles, both bearing b-sides unavailable elsewhere - "B"/"Classic Remains"/"Alone On Piano" and "Inventory"/"This Picture".

The all-instrumental Provisionally Entitled The Singing Fish, and Not To appeared within six months of one another in 1981-82; his final solo effort, the somewhat inappropriately titled Commercial Suicide was released in Belgium only, in 1986. **(see Wire)**

Discography: A-Z (Beggars Banquet 1980); Provisionally Entitled The Singing Fish (4AD 1981); Not To (4AD 1981) Commercial Suicide (Crammed Discs 1986); It Seems (Crammed **Discs 1989).**

NINE INCH NAILS

Trent Reznor remains one of the handful of true visionaries to have emerged from the post-Reagan wasteland of 80s' America, and despite early (uncomplimentary) comparisons to **Ministry** and **Skinny Puppy**, he has consistently lived up to his own advance press. Pretty Hate Machine, NIN's 1989 debut, remains one of the most challenging albums of the last decade, and spawned the alternative hit "Head Like A Hole".

Although some doubts linger as to the Nails' lasting importance, there is no avoiding Reznors' personal impact. He has also become involved with **Pigface** with **Martin Atkins** and **William Rieflin**, and has contributed to both Ministry and the **10,000** Homo DJ's side project.

Problems with his label, TVT, caused Reznor to take a lengthy break before the release of the Broken EP; maintaining the promise (if not the diversity) of Pretty Hate Machine, Broken focuses many of the ensuing frustrations and features a cover of **Adam Ant**'s "You're So Physical". **(see Ministry, Pigface)**

Nine Inch Nails

Nitzer Ebb

Discography: Pretty Hate Machine (TVT 1989); Broken (Interscope 1992); **Fixed (Interscope 1992).**

NITZER EBB

"Every band has to start taking risks to progress. It's pretty much in the lap of the Gods whether [anyone] will understand the changes" - **Bon Harris**, 1990

Emerging from the social wasteland of Chelmsford, England, a town better-known for its proximity to London than anything else, Nitzer Ebb's debt to the German bands of the early-mid 1970s is evident from their name alone.

Formed "as a hobby" at school, Nitzer Ebb was originally a three piece, but only Harris and **Douglas McCarthy** have remained throughout the band's existence - extra members are brought on board for live work.

Taking industrial dance to its furthest, most grotesque, extremes, Nitzer Ebb are the musical equivalent of a concrete tomb, a magnificent claustrophobia through which McCarthy's vocals seep like acid.

At the same time, associations with **Depeche Mode** - recent work has shared producer **Flood**, and the two bands have toured together - have seen Nitzer Ebb embrace a considerably brighter side. That Total Age, the band's first album, delivered in abundance the promises that Isn't It Funny How Your Body Works, their debut EP, delivered; early singles, too, should be treasured, and the German compilation So Bright So Strong is a convenient point of entry for anyone without the necessary cash to battle a bitter collectors' market for copies. Indeed, these three discs alone should fill anybody's desire to hear more of Nitzer Ebb: Belief was a saddening retread of That Total Age; while Showtime could have been another band entirely. Which is probably why it became their biggest hit. Ebbhead, heralded in Alternative Press as the band's best, restores a little faith.

Discography: Isn't It Funny How Your Body Works EP (Power of Voice 1985); That Total Age (Mute 1987); So Bright So Strong (Upfront 1988); Belief (Mute 1988); Showtime (Mute 1990); As Is (Geffen 1991); Ebbhead (Geffen 1992).

NOISE UNIT see FRONT LINE ASSEMBLY

NON see Boyd Rice

GARY NUMAN

Tubeway Army was a no-better-than-average new wave band, pounding the support circuit throughout 1978 (one memorable show paired them with punk immortals **The Adverts** for a night of abuse and beer cans) with interminable speedball renditions of "White Light White Heat" when Numan first discovered the power of electronics, arguably through **Donna Summer**'s summer #1 "I Feel Love", **David Bowie**'s recent Berlin sojourn, and the national impact of **Kraftwerk**'s Trans Europe Express/Man Machine albums.

Indeed, for some time after, Numan was considered little more than a Bowie clone, an impression his self-induced androgyny and claims of mechanical bisexuality did little to alleviate. Certainly it was a long way from the Army's debut single, "That's Too Bad", and the rough guitar demos which were subsequently released as The Plan.

"Bombers", a great single which certainly did not deserve to be ignored, followed, but with Tubeway Army splintering around him, by the time the Army made their first, self-titled, album in 1978, vocalist Numan had already begun dabbling in basic electronics. Early demos from this period appear on the 1984 cassette album Gary Numan 1978-79; by 1979, Replicas found him embracing the genre wholeheartedly, as the hit single "Are Friends Electric" proved.

Rocketing to #1, "Electric" carried in its wake the Replicas album - Numan also made highly successful appearances on television's Old Grey Whistle Test and radio's John Peel Show (released a decade later as an impressive mini album), arenas which insisted that his appeal would reach far beyond the basic teenybop crowd his image - featureless blonde and black leather -seemed guaranteed to cultivate.

Gary Numan

Gary Numan

At one point, Numan threatened to indeed emulate his idol David Bowie, and there is little doubt that Bowie's own next album, Scary Monsters, owed as much to Numan as Numan himself owed to Bowie. More intriguingly, contemporary accusations that Numan's music was cold and heartless today stand revealed as false - "Are Friends Electric", the somewhat anaemic vocals aside, remains a killer cut; other Numan cuts from the era - his second #1 "Cars", "Down In The Park" and "Complex" also evince a stunning maturity which was not particularly evident at the time.

"Cars" trailed The Pleasure Principle, the first Numan album to discard the Tubeway Army tag, and towards the end of the year, Numan went on the road in Britain with one of the most extravagant shows of the post-punk period.

America succumbed to Numan's charm as swiftly as Britain - "Cars" made #3, and as the year progressed, his songs and look were covered by artists as disparate as **Robert Palmer** and **Alice Cooper**.

Like **Adam Ant**, the only contemporary artist of comparative stature, Numan's appeal was not built to last. Although Telekon gave him his third successive UK album chart topper, neither "I Die You Die", "We Are Glass" nor "This Wreckage" could repeat their predecessors' success, and Numan - perhaps unknowingly - nailed the last nail into his own coffin with his highly publicized decision to retire from live work at the end of 1980. He bade farewell with the Living Ornaments 1979-80 live album.

Little was heard from Numan during 1981; he guested (as vocalist) on former Tubeway Army guitarist **Paul Gardiner**'s "Stormtrooper In Drag" single, and another old boys gathering, Dramatis' "Love Needs No Disguise" 45, but his own next effort, "She's Got Claws", saw Numan sporting a new image, that of a 1930s gangster - and a right pratt he looked as well.

Shifting away from electronics as swiftly as he moved into them, by the time Dance was released, late in 1981, Numan's audience had split in two; the survivors were still plentiful enough to push Dance to #3 in the chart (and the even more lackluster I Assassin and Warriors sets to #8 and #12 rerspectively), but even Numan's decision to reverse his earlier no touring promise could not halt his slide. Today, his still loyal following is the prime beneficiary of Beggars Banquet's multi-reissue program of Numan artefacts, whom he left in 1983.

Paramount amongst these was 1987's startlingly modernistic remix of "Cars", undertaken by producer **Zeus B Held**; with Numan digging out his old pre-retirement threads for TV promotions, "Cars (E Reg)" gave him his first major hit in five years. Japanese fans have also seen the nine CD Asylum 1-3 series, repackaging Numan's entire Beggars Banquet catalog. The first volume, traveling between 1978-80, is especially recommended. A more reasonably sized compilation, Exhibition, is also worth picking up.

Since this time, Numan's career has remained frustrating; his collaboration with **Radio Heart**, evidenced by a trio of 1987 singles, promised much, but an album - which gave the impression that Numan was involved throughout - proved that these three tracks were his sole contribution. A lengthier period working alongside disco darling **Bill Sharpe** (of the **Shakatak** group) was more successful and fruitful, but offered little of value; while Numan's own albums have grown increasingly flaccid, often leaving the listener with the impression that he's been savaged by a dead sheep. His earlier work, however, remains crucial to the future development of the industrial-electronic scene. **(see David Bowie, Kraftwerk, Donna Summer, Synthipop)**

Discography: Tubeway Army (Beggars Banquet 1978); Replicas (Beggars Banquet 1979); The Pleasure Principle (Beggars Banquet 1979); Telekon (Beggars Banquet 1980); Living Ornaments 1979 (Beggars Banquet 1981); Living Ornaments 1980 (Beggars Banquet 1981); Living Ornaments 1979-80; (Beggars Banquet 1981); Dance (Beggars Banquet 1981); I Assassin (Beggars Banquet 1982); Warriors (Beggars Banquet 1983); Beserker (Numa 1984); White Noise (Numa 1985); The Fury (Numa 1985); Strange Charm (Numa 1986); Radio Heart (NBR 1987); Exhibition (Beggars Banquet 1987); The Peel Session (Tubeway Army Strange Fruit 1987); Metal Rhythm (Illegal 1988); Automatic (with Bill Sharpe Polydor 1989); New Anger (Illegal 1989); The Skin Mechanic (IRS 1989); The Collection (Castle 1989); Outland (IRS 1991); Machine and Soul (Numa 1992)

NUMB

Based in Vancouver, Canada, Numb is the brilliant brainchild of **Don Gordon** (ex-**Images in Vogue**), formed in 1986 as a vehicle "with which to explore and manipulate concepts of social desensitization through aural constructions".

Numb comprises an ever-changing personnel, most recently restricted to Gordon and **Shards** alone, and was described by Rockpool as "nightmare ambience... oscillating between strident minimal techno-disco and atmospheric sonic freak-outs." Elsewhere, Numb have also been credited with "industrial grunge" and "the stuff of aural nightmares".

The first Numb release, the six track Blue Light cassette, appeared in 1987; the following year brought the eponymous debut album (recorded during 1986-87). Recorded with **Dave Hall** and **Sean St. Hubb**, Numb is a masterpiece of bleak ambience, with vocals vaguely reminiscent of a latter-day **Al Jourgensen** (the music, however, is nothing like him!). The Belgian label KK's 1992 CD reissue includes two bonus cuts, "Endless Descent" and "Carcinoma Angels".

The bellowing rage of Christmeister, released to great acclaim in Europe in late 1989, did not appear domestically (through BMG) until the following year; a number of record manufacturers having refused to become involved with it, due to allegedly controversial lyrical contributions by **Blair Dobson**. (The psychotic sounding Dobson is now a member of **Dead Surf Kiss**.) A video for "Cash" proved similarly controversial. Bliss comprises a number of remixes by **Greg Reely** of **Frontline Assembly** fame. Gordon has also been involved in 1993 releases by **Lung** (the stunning Magnum Opiate debut album) and **Superconductor**.

Death on the Installment Plan is forthcoming at the time of writing. **(see Images in Vogue, Lung)**

Discography: Blue Light (Burning, 1987); Numb (Edge, 1988); Christmeister (Onslot/BMG 1990); Bliss (Onslot 1991); Death on the Installment Plan (KK 1993).

NURSE WITH WOUND

Steve Stapleton's psychotic lexicon of tape effects, improvisation, found sounds and general cacophonics has produced some of the most challenging, and a lot of the most indulgent, music of the past decade. Rooted primarily in a staunch refusal to consider anything that might veer towards melody, structure and/or a tune, Nurse With Wound has blistered its way across a lifetime of albums and EPs, any of which could be considered representative. And while the project is currently considered at a close, Stapleton having retired to a farm in Ireland... well, a lot of other people have retired as well over the years.

Early NWW is perhaps the most challenging, Stapleton literally working without the references which later critical acclaim or disapproval afford even the most disinterested artist. Compiled together on Psilotripitaka, alongside a fourth, unreleased, album (Ladies Home Tickler), they show Stapleton to be moving into areas which even the established avant-garde had yet to fully explore. Challenging, and not a little trying, the three are best heard in the dribs and drabs offered by Present the Sisters of Pataphysics, a 1988 compilation.

Working with an increasing number of guest artists (**Clint Ruin/Foetus** appears on Brained; **Current 93** on Nylon Coverin' Body Smotherin'), Stapleton's genius is held in tighter check on later NWW albums, although he never loses his capacity to annoy and confuse. A Sucked Orange, upon which each of the twenty tracks was intended to convey the impressions given by their titles, is perhaps his most successful outing, although the triple Soliloquy for Lilith, too, cannot be ignored. (**see Current 93, Foetus**)

Discography: Chance Meeting of a Sewing Machine and an Umbrella on a Dissecting Table (United Dairies 1979); To The Quiet Men From A Tiny Girl (United Dairies 1980); Merzbild Schwet (United Dairies 1980); Insect and Individual Silenced (Un ted Dairies 1981); Homotopy to Marie (United Dairies 1982); Ostranenie 1913 (Third Mind 1983); Gyllenskjold, Gigerstam and I at Rydberg's EP (Laylah 1983); Brained by Falling Masonry EP (Laylah 1984); Sylvie and Babs Hi Fi Companion (Laylah 1985); Automating I (United Dairies 1986); A Missing Sense (with **Organum** United Dairies 1986); Spiral Insana (Torso 1987); Drunk With The Old Man of the Mountains (United Dairies 1987); Alas The Madonna Does Not Function EP (United Dairies 1988); Soliloquy for Lilith (Idle Hole 1988); Nylon Coverin' Body Smotherin' (with **Current 93** United Dairies 1988); Automating 2 (United Dairies 1988); Present the Sisters of Pataphysics (Idle Hole 1988); A Sucked Orange (United Dairies 1989); Coloorta Moon EP (Idle Hole 1989); Soliloquy for Lilith 5/6 (Idle Hole 1989); Psilotripitaka (United Dairies 1990)

10,000 HOMO DJ'S see Ministry

THE ORB see KLF

ORGANIZATION see Kraftwerk, Kraut Rock

ORGANUM see Nurse With Wound

OUT OUT

Inspired by On-U Sound producer **Adrian Sherwood**, Massachusetts-based **Mark Alan Miller** issued the first Out Out album, Pepperbox Muzzle, in 1991, a collection of electronic tinkerings recorded over the previous two and a half years.

Released privately (and, in abbreviated form, reissued by Turn of the Century), featuring contributions from **Tackhead** supremo **Skip McDonald**, Pepperbox was an unhinged accumulation of influences; Finched, issued on McDonald's own label, is more focussed. **Psychic TV** guitarist **Fred Gianelli** remixed several tracks. Miller also records under the name **Vein Cage.** (**see Psychic TV**)

Discography: Pepperbox Muzzle (cassette 1991): Finched (Axis 1992)

P'O see Dome

PALAIS SCHAUMBURG

The poor relations of the revived German electronics scene of the late 1970s, Palais Schaumburg emerged in 1981 with cuts on the Lieber Zuviel als Zuwenig and Festival der Guten Taten samplers. A sel-titled album appeared later that year, but the group was never destined for immportance, even after bassist/vocalist **Holger Hiller** struck out alone with the warmly received Ich Kann Nicht Mehr Warten EP.

The band's disintegration allowed Hiller wide range, and A Bunch of Foulness was one of those rare albums which lived up to its title - Hiller's guttural German becomes as harsh and unemotional an instrument as the crashing synths and percussion employed, and both this and its follow-up, Hyperprism, range close enough to **Foetus** territory to be considered disinherited second cousins of sorts.

More recent Hiller albums, beginning with Oben im Eck, find him still cominging to grips with the new technology available to musicians; it could be some time before he recaptures his formative magic.

Selected discography: Palais Schaumburg (Phonogram 1981); Ich Kann Nicht Mehr Warten EP (Warning 1981); Wir Baunen eine Neue Stadt (**Palais Schaumburg** Kamera 1982); A Bunch of Foulness in the Pit (Cherry Red 1984); Hyperprism (Wave 1985).

Pankow

PANKOW

Pankow were born in 1981 by an intuition of **Urizio Fasolo**, leader and engine of the band, the sole element that survived the multifarious mutations that the band had to pass through.

The band got in touch with English producer **Adrian Sherwood**, well known for his works with **Depeche Mode, Cabaret Voltaire, Tackhead** and more, that immediately made himself available to mix the whole album, while for a sleeve they chose the work of Swiss artist, **Hans Rudy Giger** his creature 'Alien', and on December 1987 finally "Freiheit fuer die Sklaven" appeared on the market.

Worldwide reaction and very encouraging critics: Pankow were reviewed on most European and American magazines and sales got to unexpected figures. They are pure dynamite: they could sow madness and subversion without dirtying their hands.

PENAL COLONY

From the halls of the Inland Empire and Orange County, **Penal Colony** merge out of the ashes from well known Southern California acts **Ex-Voto** and the **Texas Vamps**. Both bands have seen maginal success through Europe in The Dark Wave scene. They combine the raw elements of Killing Joke with samples and their own original dark feel to bring it up to a fresh 90's sound.

Discography: Put Your Hands Down (Cleopatra **1994**)

PERE UBU see **Punk Rock**

PETS FOR MEAT see **Controlled Bleeding**

PIG

A founder member of **KMFDM**, **Raymond Watts** debuted Pig in 1990 with the stunning

"Never for Fun" single. Poke In The Eye followed, a harsh, electronic monster which also harked back to Watt's earlier work with **Einsturzende Neubauten** (he was their first engineer) and **Foetus** (he was a featured player on the Rife album).

"Sick City" and the Praise the Lard album confirmed Pig's vision; A Stroll in the Pork reprises "Sick City" (recorded live), plus a 1988 demo, "High Protein" and two radical renditions of **Alice Cooper**'s "Hello Hooray". **(see Einsturzende Neubauten, Foetus, KMFDM)**

Discography: Poke In The Eye (Wax Trax 1990); Praise the Lard (Concrete 1991); A Stroll In The Pork (Concrete 1992)

Chris Connelly

PIGFACE

by Jo-Ann Greene (**originally published in Goldmine magazine**)

Martin Atkins was drummer for **PiL** throughout their most meaningful period, and during the late 1980s toured with **Ministry** (he appears on the Mind Is A Terrible Thing album). His solo work, released under the **Brian Brain** psuedonym, concentrates a wicked blend of anarchic musicianship and humorous lyrics, and is naturally heavy on the percussion. The biggest drawback tends to be Atkins' naturally poor singing voice.

His first singles - "They've Got Me In The Bottle"/"I Get Pain" and "Another Million Miles"/"Personality Counts" (both Secret, 1980) paved the way for the aptly titled Unexpected Noises that same year. Two further singles, "Jive Jive"/ "Hello To The Working Classes" and the brilliant Fun People EP followed.

Following a stint in **Killing Joke**, Atkins joined Ministry for a US tour. Atkins first met **Jourgensen** back in the days of PIL, but their relationship started when the Ministry man attended a Killing Joke show in Chicago. When Jourgensen called Atkins to invite him onto the Ministry tour, Atkins had a slight dilemma; he'd also been invited to join the newly reformed **Blondie**. It took him about a nanosecond to choose between them.

Atkins explains what happened next. One night during the tour, "I looked around the stage, and I said to **Bill Rieflin**, 'Hey, we're going to do something where everybody does something else other than recreate **Ministry** songs'. Because there was this wealth of talent up there... and I don't want to sound insulting, but it was like bringing together the ten greatest chefs in the world to make a Big Mac. I wanted to bring the ten greatest chefs together and say, 'O.K., do whatever you want.'"

At the end of the tour Atkins made good on his word, heading into **Big Black** supremo **Steve Albini**'s studio with everybody left standing after the tour: **Connelly, Ogre, Reznor, Yow, Tucker, Schultz, Rieflin, Barker**, and **KMFDM**'s **En Esch**. The result was **Pigface**, whose debut EP Spoonfed Breakfast" was released on Atkins' own Invisible label.

Spoonfed Breakfast comprised five tracks: "Tonight's the Night (Little Sisters)", "Winebago Induced Tapeworm remix", "The Bushmaster", "The Bushmaster remix" and the strangely titled "War Ich Nicht Immer Ein Guter Junge? War Ich Nicht Immer Schoen Und Nett? Ich erpfluckte Niemals Eine Spinne-War Niemals Frech Und Stahl" (co-written, hardly surprisingly, by Esch), and was released on cassette and vinyl only. When the sessions spilled over

Pigface

onto a full album, Gub the EP was added to that as well. "It was kind of nice for us to do that, because it deletes Spoonfed Breakfast," says Atkins. (Cassette and CD versions of Gub also include a bonus remix of "War Ich Nicht...")

Even with the absense of Jourgensen, Atkins is quick to credit the Ministry frontman for "putting the embryo of Pigface together. His contribution to the Gub album was to come into Wax Trax while we were recording there, and drink a bottle of Bushmill. And he wears his Pigface shirt occasionally, which is nice."

With Gub in the can Atkins began to wonder if the idea would translate into a more conventional project. Even though he claims Pigface "isn't a real band, it's the framework of a real band", a slimmed down version of the group formed and toured.

The proceedings were recorded for the live album Welcome to Mexico...Asshole. Barker did not take part in the show, although Rieflin did, as well as a huge cast of extras. In fact, the line-up seemed to be constantly changing, as friends showed up at different gigs to join in the fun.

Bill Rieflin summed Pigface up best on the Lean, Juicy, Pork promotional interview album, released to coincide with Welcome To Mexico. Talking to Alternative Press magazine's Jason Pettigrew, he explained "the idea of Pigface is that the people are interchangeable. As long as they're part of a certain spirit, anybody can come and play, and certain people need not play to make it happen."

William Tucker continued, "Pigface is all about fucking up preconceived notions about what's acceptable and what's not acceptable in the realm of rock music. Pigface is about being in control of your own musical destiny...(and) having a good time on the road with your best friends."

The down side of the project, as Atkins concurs, is that logistically, Pigface is a nightmare. But the shows are not quite as chaotic as they sound: some songs and elements are rehearsed, and as for the rest...well, anything can happen. At the end of the day, Atkins had the chore of going through 30 gigs worth of tapes, and choosiung the material for "Mexico".

Chris Connelly

Even so, Atkins was unhappy with the final result, and in the end decided that

"CD music was not the format in which to represent how different Pigface is. That's why we decided to do Glitch, (the band's video release) because a picture's worth a 1,000 words. I can do a 1,000 interviews and say, 'the band changes', but once people see Glitch...." Yea, now it all makes sense.

However, Pigface have released a third album, Fook. By this time, there was no Ministry connection at all, unless you count Chris Connelly, and the album sports a cast of eighteen!

Martin Atkins

Musically, Pigface is difficult to characterize. Ranging from almost experimental No Wave to industrial, each song is distinctly different from its predessors, dependent upon composers and performers. At times difficult, awkward and precocious, it's very diversity is both its strength and failure.

More successful is **Murder Inc.**, an Atkins project that saw him reunited with former Killing Jokers, **Geordie Walker**, **Paul Raven**, and **Paul Ferguson**, along with Chris Connelly and **John Bechdel**, who had also done time in Ministry.

Their first releases in 1991, the CD EP Corpuscle, and a red vinyl 12" promo coupling two of its cuts, "Murder Inc." and "Mania", was swiftly followed by their eponymous LP. More restained and much darker than Pigface, Murder Inc. has a coherent vision that Pigface inevitably, due to it's structure, lacks.

Originally signed to Atkins' Invisible label, Murder Inc. has recently been licensed to Mechanic, and their first release is set to be a remix of the album, handled by the likes of **Justin Broderick** of **Godflesh**, **Paige Hamilton** from **Helmet**, Paul Raven and many more.

Whereas Murder Inc.'s growing popularity demanded a change to a label with industry clout, Pigface is happy at Invisible. An album that will include new material as well as remixes from Fook, should be out by spring. Atkins is determined that it will sell at EP price, "because we have our own studio, it's just my time and some tape involved in remixing. We don't have to make back $4,000 in studio time, thus we can give people a really good deal and highlight how pathetic the major labels are." **(see Killing Joke, Ministry, My Life With The Thrill Kill Kult, Public Image Ltd)**

Selected discography: MARTIN ATKINS Unexpected Noises (Secret UK 1980); Fun People EP (Secret 1981); Fun With Music! EP (Plaid 1985); Time Flies When You're Having Toast (Moving Target 1987). PIGFACE: Spoonfed Breakfast EP (Invisible 1991); Gub (Invisible 1991); Welcome To Mexico... Asshole (Invisible 1992); Lean, Juicy and Pork (Invisible 1992); Fook (Invisible 1992) MURDER INC.: Murder Inc. EP (Invisible 1992); Murder Inc (Invisible 1992)

Pink Floyd

PINK FLOYD

The archetypal British psychedelic band was formed by guitarist **Syd Barrett** in 1965, and developed its trademark improvisations from playing interminable renditions of 60s' R&B classics. Early recordings - the Piper At The Gates of Dawn album and accompanying singles - showed no more of the band's instrumental prowess than Barrett's own writing demanded; only the lengthy "Interstellar Overdrive" captures the true power of the group. (Recommended are all three different recordings now available, on Piper, the compilation Relics and the soundtrack Tonite Let's Make Love in London).

Amidst growing speculation concerning his sanity, Barrett quit early in the sessions for Saucerful of Secrets, and over the course of four more albums, Pink Floyd drifted through a series of meditative experiments, the most succesful of which are included on the Atom Heart Mother (the band's first UK #1), Meddle and Obscured By Clouds (movie soundtrack) trilogy. Saucerful and another movie excursion, More, are also interesting if unfulfilled. Ummagumma, frequently cited as their most courageous release, is also their most boring; a double album, it is divided between a repetitive live disc and an album's worth of excrutiating solo offerings. The 8-track release, incidentally, omits 3/4 of the live half. (Outside of the main band project, the finest solo composition by any Floyd member remains guitarist **Dave Gilmore**'s novelty bopper "I Want More", a surprise hit single for **Can** in 1976.)

The last truly essential Floyd album, Dark Side Of The Moon (1973), was also their most successful; talk of an album recorded entirely on household items fell through when the band realized just how difficult it would be to learn to play the spoons, and while **Roger Waters** did involve himself in the bizarre gurglings of The Body, recorded with **Ron Geeson**, the Floyd itself eventually set the controls for the heart of the bank with Wish You Were Here. **(see Can)**

Selected discography: Saucerful of Secrets (Columbia 1968); More (Columbia 1969); Atom Heart Mother (Columbia 1970); Meddle (Columbia 1971); Obscured By Clouds (Columbia 1972); Interstellar Overdrive EP (See For Miles 1990); **Dark Side Of The Moon (Columbia 1973).**

CONNIE PLANK see Holger Czukay, DAF, Einsturzende Neubauten, John Foxx, Killing Joke, Kraftwerk, Kraut Rock, Neu

IGGY POP see David Bowie, Brian Eno

POP GROUP

Half **Captain Beefheart**, half **Sex Pistols**, Bristol's Pop Group - **Simon Underwood**, **Mark Stewart** and **Gareth Sager** - emerged out of the post-punk hangover with a startlingly eclectic death dance which at the time prompted more loathing than anyone since **Suicide** made the mistake of thinking they could open for **the Clash**. Supporting darlings of eccentricity **Otway-Barrett** in 1979, the Pop Group were even barracked by the show's DJ compere.

Y, a tentative beginning for a band capable of provoking such emotions, was followed by the seminal For How Much Longer Do We Tolerate Mass Murder?, a militant album which hauled into view the advances made by bands normally considered far beyond the Pop Group's own narrow confine of friends - **the Slits**, for instance, and the highly under-rated **Doll by Doll** - as well as establishing for the Pop Group themselves a reputation far greater than the group itself. **Sonic Youth** and Michigan funkmetal favorites **Big Chief** are amongst the artists who today cite the Pop Group as a primal influence.

The seminal "We Are All Prostitutes" single earned a degree of non-musical notoriety with its b-side, a recording of the Amnesty International Report on British Army torture of Irish prisoners; also assisting the band's rise was a joint single with **the Slits**: "Where There's A Will".

The band broke up in 1980 (We Are Time is a posthumous collection of live tracks and out-takes); from their ashes emerged **Rip Rig & Panic**, **Pigbag** and **Mark Stewart**.

Of these, the latter is frequently included in Industrial round-ups, although Stewart's first excursion, Learning to Cope With Cowardice (recorded with the **Adrian Sherwood/ Tackhead-**powered **Maffia**) is closer to the dislocated reggae which the Slits engaged on their Giant Slits album.

The Veneer of Democracy, two years later, is a broader offering, comparable to early Cabaret Voltaire; this initial stage of Stewart's journey is neatly encapsulated on Mark Stewart and the Maffia (Upside 1986) a US only compilation drawn from the two albums.

Mark Stewart and Metatron continue in a similar vein to Veneer of Democracy, but in a less harsh (if not bombastic) fashion. Innovative if not wholly enjoyable, Stewart remains one of the most important figures on the scene. (**see Punk Rock, Adrian Sherwood, Tackhead**)

Rozz Williams circa 1984

Discography: THE POP GROUP Y (Radar 1979); For How Much Longer Do We Tolrate Mass Murder? (Y 1980); We Are Time (Y 1980)
MARK STEWART Learning to Cope With Cowardice (with the **Maffia**; Plexus 1983); As The Veneer of Democracy Starts To Fade (Mute 1985); Mark Stewart (Mute 1987); Metatron (Mute 1990)

PREMATURE EJACULATION

"I don't even know if I'd consider us Industrial.... We're basically just a couple of insane people who make a lot of noise." **Rozz Williams**, 1992

Rozz Williams erupted out of the LA underground in the late 1970s, the 16 year old figurehead of **Christian Death**. Starting life as a straightforward punk band, by the time of their first album, Only Theatre of Pain, in 1982, Christian Death had metamorphosed into a shimmering bat-black demon, both anticipating and advancing the Gothic scene then bursting through in Britain.

Over the next two years, constant live work perfected Christian Death's unique vision of B-Movie snuff rock. But their second album, Catastrophe Ballet was to be their last. Christian Death shattered on the eve of a major European tour, leaving Williams with nothing more than the name, some songs, and an irate LA-based trio called **Pompeii 99**, who had been scheduled to support the Death on the road. It was this band, with Williams on vocals, which was to carry Christian Death into the second half of the decade.

The union did not last. Shortly after the release of Ashes in 1985, a disillusioned Williams quit. And while the rest of the new-look Death soldiered on, a state of bedraggled happenstance which survives to this very day, Williams turned his attention to PE, an outfit which he first worked with as a part-time project in 1981.

In an age when Performance Art had from necessity exploded far beyond its original confines, into an arena where only the vilest was valid, the original Premature Ejaculation lost little time in making their mark. Williams remembers one show where his partner started eating - and simultaneously regurgitating - a very dead cat he had found.

"I've always been a great believer in provoking a reaction," says Williams. "I would rather make a record that is loathed by ten people, than one which is ignored by everyone; I'd rather play a show which the audience walks out of en masse, than one they just stand around and talk through."

Very few people stood or talked through a Premature Ejaculation show, and the group swiftly found itself unable to get a booking anywhere in L.A. Concentrating instead on creating taped soundscapes, it was this area which Williams returned to, when he and **Chuck Collison** formed **the Happiest Place on Earth**.

Premature Ejaculation was reborn late in the year, when the group - whose past excesses had still only partially been forgiven and forgotten - found itself with the opportunity to work live once again. The duo also resumed recording, recreating the visual excesses of their live work in sound, a brilliant potpourri which frequently defies the often conventional instrumental line up of guitar, keyboards, samplers and percussion. Innovative to the point of savage futurism, PE's stop-start recorded career remains the ultimate test of the liberated mind.

Williams' next outing, **Shadow Project** (formed with **Superheroines** guitarist Eva O), continued his interest in fusing music with art, and like **Christian Death**, Goth with industrial noise. Throughout the Project's two albums, an eponymous debut and the colossal Dreams for the Dying, Williams' refusal to be bound by even the most elementary rules of rock has lifted him far above the run of the mill Goth scene, until even conventional songs (the classic "Static Jesus" included) appear to be taking chances far beyond their audience's requirements.

Not content, however, with this one finely-tuned string to his bow, 1991 also saw Williams re-invent Christian Death, taking to the road with a brand new line-up and, earlier this year, returning to the studio to cut The Iron Mask for Cleopatra Records. A well-wrought collection of Christian Death's finest songs brought savagely up to date, he describes it as the fruition of the vision which the original Christian Death were never to execute fully.

Like **David Bowie**, his most celebrated influence, Williams has long since proven himself capable of transcending whichever genre he might be dropped into; unlike Bowie, however, there are no traces of dilettante transcience to his work. As much as any act to emerge from the cultural terrorism of the 1980s, Rozz Williams both earned and deserved the otherwise over-used tag of genius.
(**see Goth**)

Discography: CHRISTIAN DEATH: Only Theatre of Pain (Frontier 1982); Deathwish EP (L'Invitation au Suicide 1984); Catastrophe Ballet (L'Invitation 1984); Ashes (Normal 1985); The Decomposition of Violets (ROIR 1987); The Iron Mask (Cleopatra 1992); Every King A Bastard Son (**Rozz Williams** Cleopatra 1992); The Path Of Sorrows (Cleopatra 1993).
PREMATURE EJACULATION: Assertive Discipline (Anomalous 1987); Death Cultures vols 1/2 (Happiest Sounds on Earth 1987); Anaesthesia (Dark Vinyl 1993); Blood Told In Spine (Gymnastics 1993)
SHADOW PROJECT: Shadow Project (Triple X 1990); Dreams for the Dying (Triple X 1992).

INTERVIEW
PREMATURE EJACULATION/ROZZ WILLIAMS

The first Premature Ejaculation shows, in L.A. in 1981, excited considerable negative attention for their attention to horrific detail - the dead cat and so on. Is this why the band disappeared for view for so long (four years)?
Partly. We did a couple of shows which weren't very well accepted, and it became very difficult for us to find any more gigs after that. We went on experimenting as far as making tapes ourselves, and doing things like that, but that was essentially the end.

Then you regrouped with Chuck Collison in 1985....
I met up with Chuck and we began working under the name The Happiest Place on Earth, a street performance installation. Then I decided to delve into making soundscapes again, so we revived PE and started doing shows again.

Had people forgiven you by that time? Or forgotten?
It was strange, because a lot of people had forgotten PE; they mistook the PE shows for

Christian Death! But they had not forgiven us, which is why we started just doing things with tapes and so forth. We weren't too sure about doing shows, until a friend, who ran a club called the Crypt, asked us to come down.

Did she know what to expect?
We warned her. She said 'why don't you go ahead and put on this show?', so we said 'okay... hopefully you'll be ready for it, and hopefully other people will."

What happened?
We did the show and people came expecting to hear "Romeo's Distress", and to see me clad in black, and instead they were assaulted by various noises and video images and meat flying, and eyeballs being thrown at them and so forth. A lot of them were not that keen - quite a few came up to me that evening rather upset and disturbed.

Which of course was the point?
If the ad you are seeing says Premature Ejaculation, then it is not Christian Death or Shadow Project. It is Premature Ejaculation, and we have our own set of fans or whatever you would call them, and our own set of rules. People weren't that thrilled by it, but we continued with it because no, our whole point isn't to thrill people, or necessarily entertain them. It is basically to visually and aurally assault them.

What instrumentation do you use on stage?
Various samplers, metal pipes, piece of meat which we mike up, anything we can find. I try to veer away from guitar work, because I bought so many guitars to use at PE shows and by the end of the evening they'd be completely destroyed. We don't like to limit ourselves in any way. We did one performance under a freeway underpass and there was a junkyard next to it, we did by invitation only, so there were thirty or forty of our friends there, and we just began to make sounds under there.

How does this translate onto record?
The tapes and records are usually done in a more conventional manner, although I learned early on, when we weren't able to find clubs to play at, that you can manipulate a lot soundwise, just with a regular tape recorder and a tin can. You can do quite a bit, so we just try to keep it open and do as much as we can with what we have, and it goes from extemely intricate patterns and extremely complicated material, to very minimalistic stuff.

Do you find your records have a different audience to the live work?
There is a cross-over now. When I toured with Shadow Project (October 1992), a lot of people were asking me about PE and where they can get the material that's available. It is kinda nice to see people know the difference now.

How do you and Chuck prepare yourselves for PE?
I'll give Chuck a call, or he'll give me a call, and we can hear it in one another's voice, when we're speaking to each other, 'it's time to do another show'. Its almost a release mechanism; we can only take so much of what we see around us and sometimes it gets to us and we decide we're gonna spit this garbage right back out, distort it and maybe make it a little clearer for people to see.

Do you succeed?
I think a lot of people don't really get the point but that's okay; if they come away with an idea about it, and maybe see things in a different light, better or worse, then that's good, and if they don't, if they come away scratching their heads and wondering 'what did that mean?', well that's okay as well, because obviously they're not working with too much to begin with!

It must be difficult to have coherent thoughts when there's eyeballs and pieces of meat flying at you!
I suppose it can be a bit much at times, but as I said, in the same way as our soundscapes vary from one extreme to another, our shows do the same. Sometimes they're very subtle in their approach, and sometimes they're very assaultive. It depends on the mood we're in.

We do have a very serious message that we're trying to get across, but it's very difficult sometimes because of the way it's presented. A lot of people turn away from it straight away, because of the graphics we use, on our records or at the shows, and they don't really give us a chance.

Couldn't you modify the images without losing the impact of the performance?
Not really. But it's okay, because we're not in this to force ourselves on people. In fact, that is probably our most important point, that you have to realize you cannot allow yourself to be forced into any situation that you don't want.

PRESSURE COMPANY see **Cabaret Voltaire, Tiller Boys**

Pressurehed

PRESSUREHED

Tommy Johnston and **Len Del Rio** formed LA's Pressurehed in 1988, calling upon the spirits of **Wire**, **Hawkwind** and **Chrome** and swiftly winning the support of local radio station KXLU, for whom the duo made its live debut.

Joined by bassist Marc Collignon, Pressurehed have since toured with Hawkwind, **Sex Gang Children** and **Babyland**, all allowing the band to further hone their modernistic vision of space rock; Wired for Sound, in 1991 was an immediate success, its limited pressing swiftly selling out, and prompting Flipside magazine to include a free (and now highly sought after) Pressurehed flexi in its March 1992 issue. Infradrone combines new versions of old material with more recent studio recordings.

Discography: Wired For Sound (Brain Squid 1991); Infradrone (Cleopatra 1992)

PROTHESE see **Front 242**

PSYCHIC TV

Genesis P-Orridge and new wife **Paula**, having left **Throbbing Gristle** behind them, did not waste in time in continuing their musical career, this time in the guise of Psychic TV.

On paper a more musical collaboration - **Alex Fergusson** of post-punk noisies' **Altemative TV** (he later quit to form **Coil**) joins **TG**'s **Peter Christopherson** in the new group's ranks - PTV actually continued very much from where Gristle left off, wading knee deep in murky waters whose often superficial commercial leanings (a good half of the debut Force the Hand of Chance falls into that category) are suddenly shattered by bursts of atonal noise. Psychick TV Themes, which accompanied Force, continues this pattern over a series of instrumentals ranging from the conventional to the hopelessly obscure - a description which also fits the group's choice of musical instruments. There is certainly a **Current 93** influence at play here.

Psychic TV

"Hymn 23", from Dreams Less Sweet, is another example of PTV's uncanny knack for creating blistering pop music; indeed, two years later another such piece, "Godstar" (dedicated to/written about the late **Rolling Stone**, **Brian Jones**) landed PTV in the UK chart. Sadly, history does not report what a nation of smiling pop kids thought of the remainder of PTV's catalog.

Another bastion of establishment normalcy was breached when PTV bulldozed its way into the Guiness Book of Records, courtesy of a mindboggling collection of live albums, one a month over anything up to 23 months. Not all appeared in the same territories, but 14 British albums in 18 months was enough for the record keepers. Each, incidentally, was named after the city in which it was recorded, and is a distinctly separate series to PTV's other live endeavors - New York Scum Haters (the Danceteria, 11/83); two volumes of Berlin Atonal (one shared with US born percussionist **Z'ev**, the other with **La Loora**); Those Who Do Not (Iceland 1985) and so on.

Psychic TV

Another PTV hobby has been the creation of pseudonyms, whose work is included alongside PTV on mock compilation albums: Jack The Tab sees P-Orridge smilingly insinuating himself into Britain's naive acid scene, a theater which was perfect for his maneuverings. Arguably, this album and the Tekno Acid Beat volumes catch PTV at their consistent best.

P-Orridge is currently based in the U.S., having been forced out of the U.K. early in 1992. Recent projects include work with acid guru **Timothy Leary**, and a projected union with **Pigface**.

Fred Gianelli, who joined the band in 1984, has also maintained a fairly high solo profile, debuting in 1984 with the **Turning Shrines** project, whose self-titled album appeared on the band's own Temple label. He later recorded a cover of **Hendrix**'s "Are You

Experienced", under the name **Caresse and Sickmob**, with **Genesis**' daughter **Caresse**. 1991 brought the electronic dance-oriented Fred album (Wax Trax), featuring cameos from **Thrill Kill Kult**'s **Groovie Mann**. Gianelli has also worked with **Out Out** (**see Coil, Soft Cell, My Life With The Thrill Kill Kult, Out Out, Pigface, Throbbing Gristle, Z'ev**)

Selected discography: Force the Hand of Chance (Some Bizzare 1982); Dreams Less Sweet (Some Bizzare 1983); New York Scum Haters (Temple 1984); Berlin Atonal vols 1/2 (albums shared with **Z'ev, La Loora** Atonal 1984); 25 December 1984 (Temple 1985); Those Who Do Not (Gramm 1985); Themes 1/2 (Temple 1985); Mouth of the Night (Temple 1985); Live In... (up to 23 volumes, Temple 1986-87); Themes 3 (Temple 1987); The Yellow Album (Temple 1988); Live at Mardi Gras (Temple 1988); Jack The Tab: Acid Tablets volume one (Castalia 1988); Allegory and Self (Revolver 1988); Live at Thee Circus (Temple 1988); Tekno Acid Beat vols 1/2 (Temple 1989); Live at thee Ritz (Temple 1989); Kondole/Copycat (Temple 1989); Towards Thee Infinite Beat (Wax Trax 1990); Beyond Thee Infinite Beat (Wax Trax 1990).

Psychopomps

PSYCHOPOMPS

Another discovery from Denmark. The Psychopomps target the ultra hard electro fraction, a fast, noisy & hymnic synthesis of Ebm, Industrial and Metal on their debut album Assasins DK United. Co-produced by Leætherstrip's Claus Larsen.

Selected discography: Assasins United (Zoth Ommog 1991); Mordek (Slop Pail 1992); Godshit (Zoth Ommog 1992); Pro Death Ravers (Cleopatra 1993).

PTP see Ministry

PUBLIC IMAGE LIMITED

The very nature of PiL has been to keep moving, a succession of scurrying explorations which have seen former **Sex Pistols** frontman **John Lydon** investigate everything from brainless pop ("Rise", "Don't Ask Me") to 70s retro-progressive (That What Is Not) to suicidal dance music - the first three studio albums.

Line-ups have remained fluid at best, and arguably PiL's best work was that which paired Lydon with bassist **Jah Wobble**, guitarist **Keith Levene** and drummer **Jim Walker**.

Public Image Ltd.

Lydon's love of **Can** and **Peter Hammill** was no secret even at the height of the Pistols; by the time Public Image was released, a year after Lydon quit the Pistols, the publicity machine had already stoked public anticipation to fever pitch. It was rewarded with a dense, but often likeable, collection of songs which veered between the brilliantly fulfilled ("Public Image", "Annalisa") and the demonically self-indulgent ("Fodderstompf"), over which Lydon's patented yowling cut broad swathes through both the music and the listener's patience.

(This principle was continued with PiL's in concert debut, in London on Christmas Day 1978. With no public transport to rely upon, many fans walked miles to and from the Rainbow Theatre, for the sake of a 30 minute set which included two lengthy renditions of "Public Image" itself.)

As if to acknowledge Public Image's formative state, PiL rushed out Metal Box, a metal tin into which were compressed three 12" singles. Faltering only in that PiL's finest moment ever, the 12" mix of "Death Disco", was replaced here by a much abbreviated rendition called "Swan Lake", Metal Box remains an epochal album, both in terms of its packaging and its music. Certainly Lydon has never come close to scaling its heights in the years since, and the set remains the listener's only opportunity to hear PiL at their full potential, probing, pushing and powerdriving their way past both public and critical expectations. (Metal Box was subsequently repackaged in conventional two Lp format as Second Edition.)

After so awesome a release, none begrudged PiL a lay-off, during which the fairly awful live album Paris au Printemps was released. Walker had by now been replaced by **Martin Atkins**; Wobble, however, quit before The Flowers of Romance, a circumstance which did not faze PiL in the slightest. The new album was recorded without a bass player, and as a result it slips and slobbers around the drums alone, a disorienting effort which remains the peak (experimentally if not in terms of listening pleasure) of Lydon's obsession with anti-pop themes.

Certainly by the time Levene and co quit, early on in the This Is What You Want sessions, PiL had switched their focus towards more tuneful excursions - "This Is Not A Love Song" brought the band a major 1984 hit single, in the wake of which Lydon assembled a brand new group for a Japanese tour. Live in Tokyo, released as two 12" singles, captures the ensuing shambles. For a picture of what might have been, Commercial Zone is Keith Levene's own vision of This Is What You Want, compiled from early session tapes - and with his guitar parts intact (Lydon wiped them from the authorized version of the album).

Album (also released as Cassette and Compact Disc) continued the lightening of PiL's tone, and includes contributions from, amongst many guest stars, former **Cream** drummer **Ginger Baker**. (Japanese pressings of the album credit all musicians involved; UK/US sets dispense with such niceties.) Happy and 9 are similarly lightweight, characterized only by the strength of some of the songs. But hey, people don't listen to PiL for songs, do they? Apparently they do - The Greatest Hits, So Far was a startlingly powerful collection, bringing together all PiL's singles to date, including several which had seemed unlistenably coy upon release. A word of warning - the version of "This Is Not A Love Song" bears little resemblance to the spartan rant which was originally released. The curse of the remix.

That What is Not was PiL's final album for Virgin, Lydon's home since 1977. Notable for a sample of the Pistols' "God Save The Queen", churning into view at the close of "Acid Drops", the record both confirms and rejects Lydon's role as the clown prince of punk, and was a fitting preface to the storm of sprurious Sex Pistols to Reform rumors which kept the UK press hopping as 1992 faded away. (**see Martin Atkins, Punk Rock, Van Der Graaf Generator**)

Discography: Public Image (Virgin 1978); Metal Box (Virgin 1979); Paris au Printemps (Virgin 1980);

The Flowers of Romance (WB 1981); Live in Tokyo (Virgin 1983); This Is What You Want... This Is What You Get (Elektra 1984); Commerial Zone (PiL 1984); Album/Cassette/Compact Disc (Elektra 1986); Happy? (Virgin 1987); 9 (Virgin 1989); Greatest Hits, So Far (Virgin 1990); That What Is Not (Virgin 1992).

PUNK ROCK

Punk's willingness to experiment both within and without its own parameters, belied though it was by the popular conception of the music as a simple three chord napalm storm, was the one asset which allowed its message to thrive even as its medium fell into disarray.

The likes of **Cabaret Voltaire**, **Throbbing Gristle** and **the Pop Group** all had their roots in the furnace of musical liberation preached by punk; there was room inside for everyone, although with a lexicon of past glories which included the likes of **Captain Beefheart**, the **Mothers of Invention** and the **Velvet Underground**, the biggest surprise was that anyone was surprised.

The release, that summer of 1977, of **Donna Summer**'s seminal "I Feel Love" and **Kraftwerk**'s Trans Europe Express, added urgency to their strivings; more importantly, however, they threw open new boundaries of musical expression. The history of Punk's transition into the more egalitarian New Wave can thus be reckoned in terms pre- and post- these vital releases.

Thus **XTC**'s first jerks of dementia; thus **Wire**'s speedmetal translation of a **Ramones-**ridden **Kraftwerk**; thus the unacknowledged majesty of **Rikki & The Last Days of Earth**. **Daniel Miller**'s Mute catalog grew directly from the movement's D-I-Y roots; Miller himself accurately prophecised that the synthesizer would eventually become the punk rock instrument. **Sex Pistol Johnny Rotten** publicly proclaimed his love of **Can** and **Peter Hammill**, and later formed **Public Image Limited**: their Metal Box remains a keystone in the industrial wall.

Alternative TV, formed by former fanzine editor **Mark Perry** was the first indication that punk and the avant-garde could mingle -drawing their inspiration wholly from Throbbing Gristle and co, Perry presented songs of considerable complexity, marred only by the group's inability to play them to the same convoluted degree that they demanded; even the best of ATV - their The Image Has Cracked debut, the "Love Lies Limp" 45 - gives the impression of a regular pop group trying to play "Willy The Pimp".

The Fall offered a similarly convincing stab at moving into a new age (but definitely not the New Age), a Manchester-based outfit whose apparent loathing of the paraphernalia which their very existence forced them to adopt (gigs, records, interviews, the works) created a wonderfully jarring sound within which **Mark E Smith** raged like an open plague pit. More than a decade later their 1977 debut (the compulsive "Bingo Master's Breakout" 45), Alternative Press magazine was still describing them as "so incredibly awful that it's amazing they have a record deal." In actual fact, the Fall had never been without one!

Of lesser merit, but still worthy of note were **Spherical Object**, **This Heat** and **The Passage**, more or less improvisational acts whose absorption into punk came about simply because there was nowhere else they could move.

Even deeper within the recesses of punk's magnaminity, the **Spontaneous Music Ensemble** - a floating co-operative led by the likes of **Nigel Smith**, **Tony Oxley** and **Tony Braxton** - offered total musical anarchy to those brave enough to wander into their performances.

The SME's use of both "found" and adopted instruments found few takers outside of their own circle: even acknowledged US powers such as **Pere Ubu**, whose earliest releases at least were a thunderous roar of apocalyptic industrial imagery, baulked at such tactics, and it was ironic that perhaps the nearest the punk mainstream was to come to the Ensemble's refreshing noisesomeness was via **Daevid Allen**, leader of hippy heroes **Gong**, and now touring with **Here and Now**.

Frequently linking up with ATV amongst others (the two bands share a 1978 live album), **Here and Now** opened the door for another unlikely ghost of rock'n'roll Christmases past, **Nik Turner**, ex-**Hawkwind** saxophonist. His **Sphynx**, drifting into archaic ambience, was a regular sight at free festivals throughout 1977-78, and while the band's accustomed costume of dense bandaging precluded recognition, the group's ranks at one time included former **Roxy Music/Nice** guitarist **Davey O'List**. The band's sole album, 1978's eerie Xitintoday, has lately been rerecorded, for release as Sphynx (Cleopatra).

Returning to the punk mainstream, Mancunians **Buzzcocks** and **Magazine** both flirted with electronic/pre-industrial sounds, the former through vocalist **Pete Shelley's Tiller Boys** and Sky Yen projects. **Siouxsie & The Banshees**, too, were certainly to have a say in future maneuverings, the band's first two albums (The Scream and Join Hands) sharing a claustrophobic metallic quality which clattered, ground and ultimately shattered in its often robotic purity. "Metal Postcard", despite its obvious title, was to have a considerable bearing on the more commercial end of what became industrial music; while "The Lord's Prayer" impresses simply because of its unweildy delivery.

The Banshees' influence on the later Goth scene notwithstanding, the band also pioneered a growing interest in primitive technologies, notably the use of modern instrumentation to recreate primal rhythms and beats. **The Slits**, and less notably **the Raincoats** (both of whom were to spend some time on the Pop Group's Y label) took this early dementia several steps further, the former graduating from a harsh reggae-influenced racket (Cut) to the densely textured Return of the Giant Slits, an almost impenetrable collection of sound and rhythm. Non-conformist to the point of self-destruction, the Slits were seething, tumultuous, and ultimately doomed. (**see Cabaret Voltaire, Robert Calvert, Goth, Kraftwerk, Daniel Miller, Pop Group, Public Image Ltd., Donna Summer, Throbbing Gristle, Tiller Boys, Van Der Graaf Generator, Velvet Underground, Wire, XTC**)

ERIC RANDOM see Tiller Boys

LOU REED see Metal Machine Music, Velvet Underground

RENALDO & THE LOAF see the Residents

RENEGADE SOUNDWAVE

A coupling of three DJs and party promoters, this London trio debuted with the Biting My Nails EP in 1986. Featuring reggae dub, the Renegades moved into more industrial-minded experimentation with Soundclash in 1989.

RSW in Dub saw the band drop the vocals, but their schizophrenia continued, as the songs swung from dub to industrial dance. Critics preferred their 12" singles (the Thunder EP is recommended), which were more focused than any of the albums.

Discography: Biting My Nails EP (Mute 1988); Soundclash (Mute 1989); RSW in Dub (Mute 1990); Thunder EP (Mute 1990).

ROBERT RENTAL

Rental seems to have disappeared from view, but in the early 1980s his collaborations with **Daniel Miller** set the pace for subsequent electronic developments both in Britain and the US. "Paralysis"/"ACC" was his debut, a frightening blend of noise and rhythm which was nevertheless surpassed by Live, recorded with Miller at West Runton Pavilion in March, 1979.

Rental also collaborated with **Thomas Leer** on The Bridge, an album of industrial ambience upon which furious guitar attack ("Monochrome Day") vied for superiority with the more **Cabaret-**esque pulsings of "Interferon" etc. Released on **Throbbing Gristle**'s label, The Bridge reappeared in 1992 on a Mute CD.

A single on Mute, "Double Heart"/"On Location", maintained Rental's reputation, but since that time he has apparently remained silent. **(see Thomas Leer, Daniel Miller)**

Discography: The Bridge (with **Thomas Leer** Industrial 1979): Live (with the **Normal** Rough Trade 1980)

THE RESIDENTS

Few bands have ever succeeded in maintaining their personal integrity at the same time as maintaining a high, albeit underground, presence in the minds of their fans and collectors. But the Residents, over two decades, have done just that - they are an enigma which is responsible for some of the most recognizable, and influential music of the era.

The Residents

Formed, they claim, in the Louisiana swamps in the early 1960s, the Residents began - and have continued - making music purely for their own enjoyment. It was 1970 before they "released" their first tapes, the albums Rusty Coathangers for the Doctor and The Ballad of Stuffed Trigger.

At this point the band was unnamed; they became the Residents after a 40 track tape was sent to Warner Brothers in the hope that any label enterprising enough to sign **Captain Beefheart** might have room in its heart for more of the same. It didn't, and the tape was mailed back to Louisiana, care of the residents. The name stuck.

Baby Sex, another tape, appeared in November, 1971, and featured contributions from guitarist **Snakefinger** and London poet/saxophonist **Nigel Senada**, recorded at an October 1971 Residents concert (only their second in four years) in San Francisco, to whence they had recently relocated. The following year, the Residents formed their own record label, Ralph, and released their vinyl debut, the Santa Dog double single. A limited edition of 500, the records were packaged with a Christmas card and various typically obscure promotional material. The single was not, in fact, credited to the Residents; rather, the four tracks were gifted to various non-existent acts, such as **Arf & Omega** and the **College Walkers**.

Using their own mono four track studio, the Residents spent 1973 recording what was to become their debut album, Meet The Residents, packaged in a disfigured parody of the **Beatles** similarly titled album. A promotional 7" flexidisc was also pressed for inclusion in the Toronto-based File fanzine.

A remixed version of Meet The Residents followed, in a newly designed back sleeve - replacing the Beatles' faces with three prawns and a starfish. CDs of the album restore the original artwork, and add the Santa Dog single.

Work on the next Residents' album, which appeared in 1978 as Not Available, was abandoned when the Third Reich and Roll concept, a collection of malformed 60s pop classics suggested itself. (Not Available was consigned to the vault under the terms of the Residents' own Theory of Obscurity, under which the music could not be made public until its creators had forgotten it existed.) Third Reich, a magnificent album marred only by the petty controversies which surrounded the Nazi regalia with which its sleeve was decorated, was released in 1976 with many of the ideas extended across the two sides of this album reappearing in brilliantly condensed form as the "Satisfaction" single.

Now the best known unknowns operating in the field of the avant-garde, the Residents moved into the punk era with the Babyfingers EP (January 1977), comprising one side of the projected Tourniquet of Roses album. The remainder of the three-sided disc appeared as the Fingerprince album a month later. The band also contributed the two year old "Whoopy Snorp" to the LA Free Music Society's Blorp Esette compilation (sleeve design by Captain Beefheart); released another single, the bizarre The Beatles Play The Residents and The Residents Play The Beatles set, which offered a collage of Fab Four songs weaving in and out of the protracted dying chord which closes the Sgt Pepper album; released another Meet The Residents remix, and finally, produced the Residents Radio Special promo, which brought together sundry odd singles, out-takes and album tracks.

1978 was equally hectic, with the acclaimed Duck Stab EP (later repackaged alongside a second set, Buster and Glen); guitarist Snakefinger's debut single "The Spot"; the belated release of Not Available; and a repackaging of Santa Dog's opening cut, "Fire", the original version presented at the wrong speed alongside a rerecorded one. (The band's fan club, Uncle Willy's Eyeball Buddies, later released a CD featuring four different versions of each of the original Santa Dog cuts.) The band also spent some time preparing the first Residents compilation, Please Don't Steal It, released as a promo only by Ralph, but granted a full release in the UK by Virgin as Nibbles.

Eskimo, the first all-new Residents album since Fingerprince, appeared in the new year. It was compiled from notes sent to the band by Nigel Senada, who had spent a lengthy period living at the North Pole, and took the Residents' sound beyond the playful cut-ups of old and into truly avant-garde territory. Indeed, with the exception of the aptly titled Commercial Album (1980), which comprised forty one minute pop songs recorded with **Fred Frith, Lene Lovich** and **XTC** mainstay **Andy Partridge** amongst others, the Residents were now delving seriously into the realms of murky musical obscurity, overcoming the last vestiges of their earlier "novelty" appeal with the so-called Mole trilogy.

A conceptual offering, whose weak story was only compounded by the sheer obstinacy of the music, the saga was spread out over two studio albums, Mark of the Mole and The Tune of Two Cities (1981/82), an authorized live-at-the-Roxy bootleg of The Mole Show, and a fourth collection, Intermission, which contained what the band accurately described as "extraneous" music from the show. (Intermissions was later added as a bonus to the Mark of the Mole CD; Tunes is boosted with the inclusion of three out-takes.) There later followed a fifth volume, Assorted Secrets (1984), featuring one side of live Mole songs; and a sixth, The Big Bubble (Part Four of the Mole Trilogy) appeared belatedly in 1985, long after the rest of the world had forgotten the entire debacle (Mole completists shouldn't miss the seventh in the series, The Mole Show Live in Holland). Finally, several live-in-the-studio versions of Two Cities tracks were included on the Stranger Than Supper compilation.

The notion that the Residents might finally be faltering was given further weight by their scrappy collaboration with **Renaldo and the Loaf**, Time In Limbo (the best of which has been appended to the Not Available CD), and reinforced all the more by the compilation Residue of the Residents, a brilliant companion to Please Don't Steal It. (Since that time, two further compilations, Heaven and Hell have been released.)

After over a decade of producing genuinely challenging music, the Residents' grip on both their original powers of pastiche and the sheer inventiveness which they had pioneered was being weakened by a host of new bands, many of whom took individual elements of the Residents' own work to the logical extremes the Residents themselves had either ignored, or not had time to pursue.

The Flying Lizards' reworking of the Motown oldie "Money" certainly owed much to the Residents' own renderings of sixties hits, but the group's own attempts to return to basics, via the ambitious Great Composers series, was largely dismal. Neither George (Gershwin) and James (Brown) nor Stars (John Philip Sousa) and Hank (Williams), released in 1984 and 1986, were much more than amusing, and it is ironic that the Residents' most successful meeting with Brown, a riotous version of "It's A Man's Man's World", was not included on the album.

Two soundtracks, to the decade old Vileness Fats movie, retitled Whatever Happened To Vileness Fats? (1984), and the more recent The Census Taker (1985) followed, together with an excellent live album, 13th Anniversary Show. Recorded in Tokyo, with Snakefinger temporarily back in the fold, the set draws upon the entire Residents' back catalog, and was followed by another sterling offering, a 12" dance mix of Ray Charles' "Hit The Road Jack". (Highlights from Vileness Fats, incidentally, can be found on the PAL TV compilation).

The Residents' greatest personal bugbear, the overblown concept, resurfaced in 1988 when Rykodisc released both God in Three Persons and a remixed selection of backing tracks, Soundtrack to God In Three Persons, but the band returned to full creative strenghth the following year with The King and Eye, a raw and often raucous rendering of **Elvis Presley**'s greatest hits and bits. It was only with this release that the band admitted to have now created a full history of American pop in three easy stages - earlier in the year they performed two new pieces, Buckaroo Blues and Black Barry in New York; repeating the performance in San Francisco and releasing the results on a single cassette, they had created the project now known as Cube-E. (UWEB - the Residents' fan club label, has also issued a studio recording of Buckaroo Blues.)

The importance of the Residents lies not so much in all that they have done over the past two decades, but in the underlying meaning behind what they have done. At a time when only Fred Frith's **Henry Cow**, Beefheart and **Frank Zappa** were truly exploring the avant-garde within a rock'n'roll context, the Residents went shooting off at angles which even they had yet to conceive of. That the band's early 1980s' trough coincided with an explosion of new musical vision from elsewhere - the Flying Lizards, **Devo**, **The Pop Group**, **Thomas Dolby** and more - only testifies further to the Residents' unparalleled influence.

Certainly trace elements of their earliest work still surfaces in both the mainstream and alternative fields, while their input into the industrial scene, again via the Flying Lizards et al can never be calculated. The Residents were taking chances at a time when most people merely wanted to keep them. We will never see their like again. (**see Fred Frith, Snakefinger, XTC**)

Discography: Rusty Coathangers for the Doctor (private 1970?); The Ballad of Stuffed Trigger (private 1970?); Baby Sex (Private 1971); Santa Dog EP (Ralph 1972); Meet The Residents (Ralph 1974); The Residents Present The Third Reich and Roll (Ralph 1975); Babyfingers EP (Ralph 1977); Fingerprince (Ralph 1977); Residents Radio Special (Ralph 1977); Duck Stab EP (Ralph 1978); Buster and Glen (Ralph 1978); Not Available (Ralph 1978); Please Don't Steal It/Nibbles (Ralph 1978); Eskimo (Ralph 1979); Commercial Album (Ralph 1980); Diskomo EP (Ralph 1980); Mark of the Mole (Ralph 1981); The Tune of Two Cities (Ralph 1982); The Mole Show (no label 1982); Intermission (Ralph 1982); Time In Limbo (with **Renaldo & The Loaf** Ralph 1983); Residue of the Residents (Ralph 1983); George & James (Ralph 1984); Whatever Happened To Vileness Fats? (Ralph 1984); Assorted Secrets (Ralph 1984); The Big Bubble (Part Four of the Mole Trilogy) (Ralph 1985); The Census Taker (Episode 1985); 13th Anniversary Show - Live in Japan (Ralph 1986); Stars and Hank (Ralph 1986); Heaven (Rykodisc 1986); Hell (Rykodisc 1986); The Mole Show Live in Holland (Torso 1987); God in Three Persons (Rykodisc 1988); Soundtrack to God In Three Persons (Rykodisc 1988); The King and Eye (Enigma 1989); Buckaroo Blues and Black Barry (Ralph 1989); Buckaroo Blues (UWEB 1989); Stranger Than Supper (UWEB 1990); Cube E Live in Holland (Enigma 1990); Freak Show (Official 1991); Our Finest Flowers (East Side Digital 1992).

BOYD RICE

San Francisco based Rice began recording in 1977 with a series of 45s in which he stretched both his, and his audience's, expectations as far as they would go. One early single came with four holes in the center for "multi-axial rotation"; another comprised solely of **Lesley Gore** snippets, every "cry" she ever sang.

Live, he developed an equally uncompromising reputation - "in New York," he revealed, "I was one of the first people in years to actually get cat calls. People were screaming 'I want my $7.50 back!' And in Den Haag, Holland, people were real angry. They really disliked me. I had these bright lights shining in their eyes so they could barely see me - they were trying to reach up and smash the lights, but the lights were just out of their reach."

After a number of performances with avant-garde percussionist **Z'ev** at La Mamelles in San Francisco, Rice arrived in England as a guest of **Throbbing Gristle**. They staged his first ever UK show, at the London Film Makers Co-op, following which he was signed to Mute Records during **Daniel Miller**'s first flurry of **Normal-**era madness.

Rice is a pure noise merchant, creating sounds which could as easily have been one note samples from Lou Reed's Metal Machine Music as new creations. Indeed, much of his debut album was originally recorded shortly after MMM hit the streets, and like that, demands a lot more from the listener than most listeners are prepared, or even capable, of giving. Unreleased are some mixes made by **Genesis P-Orridge**; according to Gen, "Some people said it was the loudest they'e ever heard Boyd. I had it up so loud, it literally made people move. **Monte Cazazza** told me that [Boyd made] the best dance music he'd ever heard. I thought Monte was being ironic, but when I heard it the first time I understood. It was so physical."

The Black Album (reissued by Mute as Boyd Rice) was playable at any speed without any noticeable difference ("so you get four times as much for your [money]"); it was followed by the multi- axial single "Knive Ladder"/"Mode of Infection" (also reissued by Mute). This was the first release under Boyd's **Non** psuedonym, and Rice himself drilled the extra holes in the record.

Non also performed live - Physical Evidence, contains a number of concert recordings, although none feature audience reactions as radical as Rice had hitherto enjoyed. (The A-side of the Mute single, "Rise"/"Out Out Out" was also recorded live.) Pagan Muzak, comprised wholly of loop grooves, is Non's first studio set.

Rice and Miller together recorded "Cleanliness and Order" for inclusion on the Darker Scratcher compilation; according to legend, another of the tracks, "Non Watusi", mysteriously changed sometime between the master tape leaving Rice's possession and the album being pressed. It became, Rice complained, a different song altogether.

With Mute as a regular outlet for his muse, Boyd now had longer in which to concoct fresh unpleasantries; many of these he then passed over to **Frank Tovey**, and the pair had a delightful time (three years before the album's release) recording Easy Listening, a record which proudly boasts of using no musical instruments whatsoever.

By the time of the next **Non** album, Blood & Flame, and the Music Martinis and Misanthropy sets, much of Rice's original psychosis had been overtaken by other, more tunefully-inclined, artists, as his collaborations with **Coil** in 1985 amply demonstrate. (A Non tour that same year was later captured on the limited edition Sick Tour album.) The comparative drabness of these records, however, in no way detracts from Rice's importance. **(see Coil, Daniel Miller, Throbbing Gristle, Frank Tovey, Z'ev)**

Discography: Boyd Rice (Mute 1981); Physical Evidence (Non Mute 1982); Easy Listening for the Hard of Hearing (with Frank Tovey Mute 1984); Nightmare Culture (with **Coil Laylah** 1985); Blood & Flame (Non Mute 1987); Music Martini and Misanthropy (Bad 1990); The Sick Tour (Staaltape 1992).

RIKKI & THE LAST DAYS OF EARTH see Punk Rock, Synthipop

TERRY RILEY

Californian Riley was working with **LaMonte Young** during the early 1960s (**Irmin Schmidt**, later of **Can**, was also a participant), moving from there into solo improvisational work; live, he appeared alongside sax, tape recorder and electronic keyboards ("Poppy Nogood and the Phantom Band", from his best known work, A Rainbow in Curved Air, is an example of this early material.)

In 1969 Riley (whose scholarly Keyboard Studies had just been released in France) staged a two week long performance of sine wave oscillators at a Munich art gallery; amongst the witnesses were two jazz musicians, **Ralf Hutter** and **Florian Schneider-Esleben**; they promptly set out to form **Organization**, the embryo from which **Krafftwerk** later emerged.

In 1969, the scholarly Keyboard Studies was released in France; In C and Rainbow followed, confirming both Riley's abilities and reputation - prog rock band **Curved Air** took their name from the latter album.

1971 also brought Riley's collaboration with **John Cale**, Church of Anthrax; doomed as much by Cale's inexperience as by Riley's legendary stubbornness, the album remains amongst the least interesting in his canon. Persian Surgery Dervishes shows that Riley was as capable of borrowing from rock'n'roll as he was able to contribute to it. **(see John Cale, Can, Kraftwerk, Kraut Rock, Irmin Schmidt, LaMonte Young)**

Selected discography: Keyboard Studies (BYG 1969); In C (Columbia 1970); Rainbow in Curved Air (Columbia 1971); Church of Anthrax (with **John Cale** Columbia 1971); Persian Surgery Dervishes (Shandar 1973); LaMonte Young V (recorded 1964-81, 50 Skidillion Watts 1991); LaMonte Young XI (recorded 1989-91, 50 Skidillion Watts 1991).

RINGTAILED SNORTER see X MARKS THE PEDWALK

ROXY MUSIC see Brian Eno

CLINT RUIN see Foetus

SAVAGE REPUBLIC

The unique use of two bass guitars set LA's Savage Republic aside even before they adopted that name - as **Africa Corps**, the group recorded the immense (and immensely noisy) Tragic Figures album, becoming Savage Republic to avoid being lumped in with the Nazi Surf Punk fad then sweeping the California scene.

It was a smart move - the dense metallic drone which was the republic's calling card owed little to anything else happening in the region at the time; even **Boyd Rice**, the area's best-known industrial avant-gardist, was working more in the UK at the time, and the group - **Jeff Long**, **Jackson Del Ray**, **Mark Erskine** and **Bruce Licher** - were to remain unsung, even after Del Ray and new recruit **Robert Loveless** formed the distinctly dissimilar **17 Pygmies**, then carried their newfound delight in bright and breezy pomp rock over to the sessions for a second Republic album, in 1983.

The strain on the original concept of the band became unbearable; when Savage Republic split up, midway through recording Jedda By The Sea, the Pygmies hijacked the sessions and completed them under their own banner, while Licher and Erskine returned to basics with a reborn Republic. Trudge, their debut EP, borrowed much from side two of Bowie's Low and Heroes excursions, the wordless vocals and mock-Middle Eastern instrumentation. A representative sampling of all that was left behind can be found on the Recordings from Live Performances retrospective.

This set concluded Savage Republic's dalliance with industrial formats: a second album, Ceremonial, was horribly lightweight, while the in concert Live Trek 1985-86 took the best of the bands earliest roar and completely reworked it, a process not unlike (say) **Al Jourgensen** recreating With Sympathy for **Ministry**'s modern audience. Like **Current 93**, a band to whom Savage Republic have occasionally been compared to, their finest excursions today are those which might be described as folk-oriented. What a long, strange trip it's been....

Selected discography: Tragic Figures (Independent 1982): Trudge EP (Play It Again Sam 1985): Recordings from Live Performances 1983-85 (Independent Project 1992)

Irwin Schmidt (foreground)

IRMIN SCHMIDT

Can's classically trained keyboard player released his first solo album, Film Musik (Spoon) in 1980. The first of four volumes, this album says much for Schmidt's post-Can career - he recorded five movie soundtracks between 1978-83, collected on these albums.

Toy Planet (Spoon, 1981) teamed Schmidt with **Bruno Spoerri** for a less than satisfying collaboration; Schmidt's own solo career did not really get underway until 1987's Music At Dusk (WEA) album, recorded with fellow ex-Canners **Michael Karoli** and **Jaki Liebezeit**. **(see also Can, Holger Czukay, Terry Riley, LaMonte Young)**

Selected discography: Film Musik (Spoon 1980); Toy Planet (Spoon, 1981); Music At Dusk (WEA 1987).

SHADOW PROJECT see Goth, Premature Ejaculation

PETE SHELLEY see Can, Punk Rock, Tiller Boys

ADRIAN SHERWOOD see Depeche Mode, KMFDM, Ministry, Out Out

SILICON TEENS see Daniel Miller

SIOUXSIE & THE BANSHEES see Goth, Punk Rock

SISTER MACHINE GUN

Combining the "ultra-heavy" handed production of label mate **Sasha Konietzko** of **KMFDM** with collective assistance from **Jim Marcus** and **Van Christie** from **Die Warzau** and **Dave Ogilvie** of **Skinny Puppy**, SMG leader **Chris Randall** has created a rather auspicious introduction that has led to many many pair of feet as well as a few chins hitting the dance floor.

Randall has been circulating in the New York club scene for a couple years, but only began "loading the gun" while playing keyboards on the road with **KMFDM** during their North American tour in early '91. After the tour, he returned to Chicago where he began working on demos with Konietzko and was quickly inked to a long term deal with Wax Trax.

Shortly after completion of Sins of the Flesh, Wax Trax was approached by Arista Records looking for an artist to contribute an "industrial" feel to the already semi-established club hit "James Brown is Dead" by L.A Style. Chris flew to Belgium where he contributed vocals and guitar to a new version of the song which was released in late '91 and has enjoyed a lengthy existence on the club charts since then.

Discography: Sins of the Flesh (1992 Wax Trax)

SISTERS OF MERCY see Goth, Suicide

SKELETON CREW see Fred Frith

SKIN CHAMBER see Controlled Bleeding

SLITS see Punk Rock

SKINNY PUPPY

Skinny Puppy

Celvin Key (ex-**Images in Vogue Kevin Crompton**) and **Nivek Ogre** (**Kevin Ogilvie**) emerged from Vancouver's strangely convoluted punk scene in 1984, a harshly avant-garde unit whose best early work took elements of the industrial scene's most telling pacesetters and ran them through a variety of distortion effects. The Remission EP, and their second album, Mind: The Perpetual Intercourse, find Skinny Puppy at their atmospheric best - danceable, but desperately morose.

The arrival of **Dwayne Goettel** during the Mind sessions, and the loss of **Bill Leeb** (to form **Frontline Assembly**) opened Skinny Puppy's eyes a little wider, and Cleanse, Fold and Manipulate is a considerably brighter

album, retaining the themes which powered the earlier sets, but allowing the music to move into wider explorations of sound. VIVIsectVI is an even greater progression, pushing Skinny Puppy in much the same direction as **Ministry** was now moving in - as **Al Jourgensen**'s presence on Rabies confirms. (Ogre repaid the favor by appearing with Ministry on stage, and also contributing to the **Pigface** splinter project. In addition, **David Ogilvie**, Skinny Puppy's long-time producer, has also worked with both **Ministry** and the **Rev Co** splinter project.)

Strangely, Rabies is the least satisfying of later Skinny Puppy albums, a (very) dry run for Too Dark Park, an apocalyptic monster which can only be listened to on sunny days with the birds singing outside. Ogre's own onstage obsession with mass-murder and dismemberment has seldom roared so loudly, although Last Rights runs close.

Shrugging off even the vestiges of rhythm and danceability which filtered through its predecessor, Last Rights was described by Alternative Press as "deeply disturbing, very discomorting and psychologically traumatic...as unforgiving as a collapsing building."

Skinny Puppy

Absent from the album was "Left Handshake", removed after legal pressure from one of the song's sampled guest, **Timothy Leary**'s former compatriots. The song was a regular in the band's 1992 live set.

A string of 12" singles, most of which avoided inclusion on the band's albums, was meanwhile winning Skinny Puppy a great many friends on the dance circuit; The 12 Inch Anthology collects together four A and B-sides on one devilish disc. Back and Forth, a cassette only release, was similarly compiled from early (1981-84) live and studio tapes, to stem a flow of bootleg material. Amongst sundry side projects, Ogre coproduced **Diatribe**'s "Nothing" E.P. (**see Diatribe, Frontline Assembly, Hilt, Ministry, Images in Vogue, Pigface**)

Selected discography: Remission EP (Nettwerk 1984): Bites (Nettwerk 1985): Mind: The Perpetual Intercourse (Nettwerk 1986): Cleanse Fold and Manipulate (Nettwerk 1987): VIVIsectVI (Nettwerk 1988): Rabies (Nettwerk 1989): Too Dark Park (Nettwerk 1990): The 12" Anthology (Nettwerk 1990): Last Rights (Nettwerk 1992): Back & Forth 2 (Nettwerk 1992)

SLAPP HAPPY

Formed by New Yorker **Peter Blegvad**, German vocalist **Dagmar Krause** and Briton **Anthony Moore**, Slapp Happy came together under the auspices of Kraut Rockers **Faust**, who provide backing on Slapp Happy's debut album Sort Of (recorded at the Germans' Wumme studio complex).

A self-titled album two years later continued the heavily improvisational sound of its predecessor, but it was in collaboration with labelmates **Henry Cow** that Slapp Happy came of age, the two albums they recorded together ranking as the best in either band's catalog.

Unfortunately, only Krause opted to remain with Cow; Blegvad returned to New York, Moore to London, from whence both embarked on tentative solo careers. To date Moore has had the greatest success. **Paul Young**'s multi-million selling No Parlez album in 1982 included a version of "Industrial Drums", the title track of moore's then-current solo album. (**see also Faust, Fred Frith**)

Discography: Sort Of (Polydor 1972): Slapp Happy (Virgin 1974): Desperate Straights (with **Henry Cow** Virgin 1974): In Praise of Learning (Virgin 1975)

SLAVESTATE see Sleep Chamber

SLEEP CHAMBER

Brainchild of Boston's **John ZeWizz**, with an ever changing group of extras, Sleep Chamber outpouring began in 1982 with the Speak in Tongues EP. Since than, the band has released an incredible body of work, spread over albums, EPs and cassettes, none of which seem to offer more than a token departure from the last (or next).

Musically, the band has an almost trancelike quality, over which ZeWizz's disinterested vocals wash with mantra-like calm. Lyrically, Sleep Chamber borders on the sexually obsessive, with often graphic descriptions thrown in to keep **Tipper Gore** and co on their toes. Hardly surprisingly, Sleep Chamber have spent their entire career being either banned or ignored.

Sleep Chamber

Ex-member **Thomas Lockyear** later developed the **Slavestate** project, working with **TKK**'s **Buck McCoy**. **(see My Life With The Thrill Kill Kult)**

Selected discography: Speak In Tongues (Inner X 1982); Sleep Chamber (Inner X 1984); Submit to Desire (Inner X 1985); Best of the Rest (Inner X 1986); Babylon EP (Inner X 1987); Spellbondage (Inner X 1987); Sharp Spikes and Spurs (Inner X 1989); Hot Pants and Wet Panties EP (Inner X 1990); Cat Woman (Inner X 1992); Secret To 23 (Inner X 1993).

SNAKEFINGER

Snakefinger is **Philip Lithman**, otherwise known for work with **Junior's Blues** and British pub rockers **Chilli Willi & the Red Hot Peppers** (positively no relation!). Lining up with the **Residents** very early in their career, he all but became a functioning xth member of that band, and that despite much of his work being several degrees saner than anything the Residents ever produced -early singles such as "The Spot" (c/w a version of the Residents' own "Smelly Tongues"), and "Kill The Great Raven"/"What Wilbur?" pay only token lip-service to the Residents, despite their obvious input (they cowrote much of his material). His mutilation of **Kratwerk**'s "The Model", for a 1979 single is positively respectful.

That track is one of several excellent efforts included on Against The Grain, Snakeinger's first solo album; Greener Pastures, which was again co-written and co-produced by the Residents, offers similar tentative duck stabs at blending normalcy with Residential weirdness. It was no surprise when he put together the **Dead Residents**, revolving around himself and **Eric Drew Feldman** ex-**Beefheart**, for the Manual of Errors album.

Having taken time away from his own career to tour Japan with the Residents (captured on their 13th Anniversary live album), Night of Desirable Objects and the in concert Live in Chicago cassette continued Snakefinger's collaboration with Feldman into a new group, the **Vestal Virgins**. Whilst retaining the semi-anarchic spirit of the originals, the latter features interesting reworkings of "The Spot", "The Model" and another first album cut, "Magic and Ecstasy".

Snakefinger died from a heart attack in July, 1987; the following year's CD Collection updates an earlier compilation, Against The Grain. **(see The Residents)**

Discography: Chewing Hides the Sound (Ralph 1980): Greener Pastures (Ralph 1981): Manual of Errors (Ralph 1982): Against The Grain (Ralph 1983): Night of Desirable Objects (Ralph 1986): Live in Chicago (Ralph 1986): The CD Collection (Ralph 1988)

SOFT CELL

The most innovative of all the so-called New Romantic bands which erupted out of Britain during the early 1980s, Soft Cell was the union of a self-confessed "horrible little boy with a lurid imagination" (**Marc Almond**) and electronic discobopper **Dave Ball** -a character who shared Almond's unlikely dream of crossing Northern Soul with **Can**.

The duo's first performances were multi-media extravagances, staged to little more than local recognition during 1980; that same year Soft Cell recorded their first demos, a five track session from which was culled the excellent four song Mutant Moments EP. ("Metro MIX" in particular is essential listening, a vicious amalgam of **Throbbing Gristle**'s twisted pop as performed by the **Velvet Underground**.) The remaining cut, "Girl with the Patent Leather Face", was contributed to an electronic compilation being put together by the fledgling Some Bizzare label.

Signing to that label, Soft Cell released the "Memorabilia" 12", a claustrophobic dance mutant whose theme - the joys of vacationing abroad - was utterly belied by the menacing instrumentation and Almond's pervertedly salacious vocals. He remains the only man who could make "Torremolinos" sound like a sexual act.

Described by journalist Chris Bohn as "**Suicide** made painless", Soft Cell continued eking out their dark vision with a medley of **Ed Cobb**'s "Tainted Love" and the **Supremes**' "Where Did Our Love Go?" It was a brilliant record, but its progress was slow - for two months the single scratched at the UK chart before finally coming to rest at the top; in the US, it took four months to reach the Top 10.

Soft Cell

Having arrived there, however, both it and Soft Cell seemed destined to stick around. "Bedsitter" and "Say Hello Wave Goodbye" both offered watered down versions of the duo's original dream, but still Soft Cell seemed an odd addition to the UK pop circus - just how odd was revealed by Non Stop Erotic Cabaret, an anxiety-whipped adolescent's sleaze-ridden journey through the underbelly of London's red light district. "Sex Dwarf", possibly the funniest comment on perversion ever committed to vinyl, was singled out for particular attention - drawn from a tabloid headline ("Sex dwarf leads 100 disco dollies to a life of vice"), it was accompanied by a video so explicit that it has still to be shown.

Maintaining their stranglehold on the Top 30 through singles like "Torch", "What?" and "Where The Heart Is" (but allowing their more adventurous notions to surface on b-sides and 12" mixes), Soft Cell's second album, The Art of Falling Apart remains the pinnacle of their career. "Martin" and an electronic medley of three Jimi Hendrix songs, originally included as a bonus 12" single with the album, the brooding "Numbers" and the frenetic title cut are the commercial side of industrial dance gone mad.

Throughout this period, Almond further cemented his underground credentials by guesting with **Coil** and **Psychic TV**; forming the **Immaculate Consumptives Non Stop Revue** with **Nick Cave**, **Lydia Lunch** and **Clint Ruin** (he and Ruin also collaborated as **Flesh Volcano**); and launching his own **Mambas** side project. Abandoned introspection running riot over songs by **Peter Hammill**, **Syd Barrett** and **Jacques Brel**, the two Mambas albums echo much of **Dagmar Krause**'s work in content and **Suicide** in delivery ("Ghost Riders" frequently appeared in the group's live set).

The third Soft Cell album, This Last Night in Sodom, was released hot on the heels of the group's farewell concerts in London. The heaviest thing Soft Cell had yet attempted, it was only partially successful, although Soul Inside, the final Cell EP, marked another high from a union which certainly deserved greater recognition during its lifetime than it was eventually afforded.

Almond went solo; Ball's activities include work with **Cabaret Voltaire** and **Stephen Mallinder**'s shortlived **Love Street**.

Non Stop Ecstatic Dancing is a custom-made collection of six dance remixes, most notable for the hit "What" and a version of "A Man Could Get Lost", hitherto available only on the scarce "Memorabilia" 7". Both The Singles 1981-85 and Memorabilia are posthumous compilations, the latter marred by some horrendous remixes and a swathe of Almond's sleaze-for-slime's-sake solo work. Seekers of the infinitely superior 12" mixes need hunt out a boxed set issued in limited quantities in 1982 and reissued on CD a decade later. **(see Cabaret Voltaire, Coil, Foetus, Goth, Stephen Mallinder, Psychic TV, Suicide, Synthipop, Throbbing Gristle)**

Discography: Mutant Moments (Big Frock 1980); Non Stop Erotic Cabaret (Some Bizzare 1981); Non Stop Ecstatic Dancing (Some Bizzare 1982); The Art of Falling Apart (Some Bizzare 1983); This Last Night in Sodom (Some Bizzare 1984); The Singles 1981-85 (Some Bizzare 1985); Memorabilia (Some Bizzare 1990).

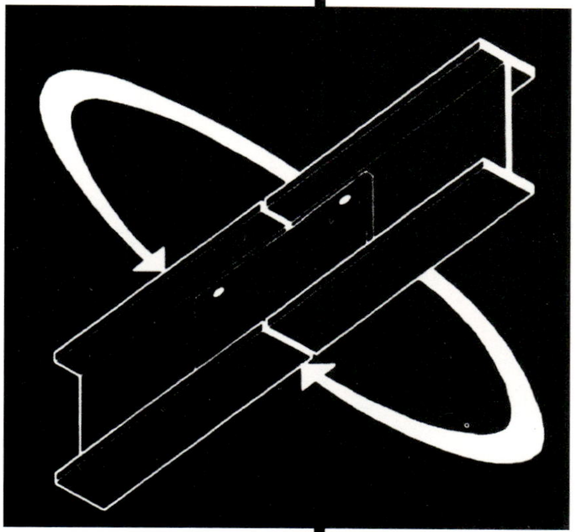

SPAHN RANCH

Comprising New Yorkers **Matt Green** and **Rob Morton**, Spahn Ranch debuted on an eponymous 1992 EP and if you like to take your industrial music neat, this pounding Leviathan will transport you to places you never knew existed. Harsh, bright and inventive are not words I bandy around every day, but they fit Spahn Ranch. This is to be followed up with their debut Collateral Damage with their new vocalist **Athan Maroulis**. Expect it mid '93.

Discography: Spahn Ranch (Cleopatra 1992); The Whip (Cleopatra 1993); Collateral Damage (Cleopatra 1993).

SPHERICAL OBJECTS see Punk Rock

SPHYNX see Robert Calvert, Punk Rock

SPK

Formed in Sydney, Australia in 1978, but subsequently relocating to London, the mysteriously named **EMSAKS** and four friends took the name **Socalist Patients Kollektiv** from a group of West German mental patients who, inspired by the Baader Meinhof Group, attempted to form a terrorist group. They died whilst making their first bombs in their hospital room.

At other times, SPK has stood for **Surgical Penis Klinik**, the name under which **Throbbing Gristle**'s Industrial label released their "Mekano"/"Slogun" single. The sleeve of this 45, incidentally, depicts a skewered penis. Live, they have eaten sheep brains raw from a recently opened skull, and - whilst demonstrating a flamethrower - set a member of their audience afire. Strangely, both events occured in San Francisco.

SPK reached their peak around 1980-81, with shows supporting Throbbing Gristle and **A Certain Ratio** in London, **Factrix** in the US. Live at the Crypt was recorded at one of the London shows, and remains the most easily absorbed of all SPK recordings. As with Throbbing Gristle, SPK's recorded career is more a by-product of other artistic aims than a be-all and end-all; their stark, tuneless recordings are similarly indicative of moods and expressions, although not the sort of moods or expressions most people would want to experience. **(see Throbbing Gristle)**

Discography: No More EP (Side Effekts 1979); Faktory EP (Side Effekts 1979); Information Overload Unit (Side Effekts 1980); Live at the Crypt (Sterile 1981); Solipsik EP (M2 1981); Leichenschrei (Thermidor 1981); Last Attempt At Paradise (Fresh 1981); From Science to Ritual (Viva 1981); Auto-da-fe (Walter Ulbricht 1982); Dekompositions EP (Side Effekts 1982).

SPONTANEOUS MUSIC ENSEMBLE see Punk Rock

STEREOTAXIC DEVICE

An LA based duo comprising ex–**Third Eye** member **George Sarah** and **Jennifer Willlams**, Stereotaxic Device debuted in 1990 with the S/T album, a sharp, often surreal collection which established the band as a force to be reckoned with, both musically and politically. The two members are fervent animal rights activists (Williams was stage producer for **Skinny Puppy's** VIVIsect tour) and their beliefs are an integral part of the band's make-up. Several venues, incredibly, have refused to rebook Stereotaxic Device until they promise not to accompany their performance with graphic videotapes of animal research work (work, incidentally, which both private and governmental researchers claim is humane. So why is it so upsetting to people?)

Further, similar, bricks and bouquets awaited the duo's second (and so far finest) album, the thoughtfully titled 100 Per Day Extinct. A stormy blend of techno dance and gloom-ridden moods, set to a frantic multi-BPM rhythm, 100 Per Day was on the receiving end of several quite spiteful reviews including one which not only lambasted the political content of the lyrics, but actually tried justifying the experiments the duo deplore.

Discography: S/T (KK 1990); 100 Per Day Extinct (KK 1992).

STEROID MAXIMUS see Foetus

MARK STEWART see Pop Group

KARLHEINZ STOCKHAUSEN

German-born composer Stockhausen began working in electronic music in 1953 from his studio in Cologne; two years later, he co-founded (with **Herman Eimert**) Die Reihe, the first magazine devoted to the field. His first major electronic composition, Gesang der Junglinge, for tape and treble, appeared in 1956.

In 1957, upon publication of Gruppen, he staged the first of his now institutionalized composition courses at Darmstadt; amongst his pupils can be included **Irmin Schmidt** and **Holger Czukay**, later of **Can**.

His growing reputation within avant-garde rock'n'roll fields led to a major resurgence of interest in his work during the 1970s: Fur Kommanden Zeiten, in 1970, proved that the competition had in no way dulled his edge. (**see Can, Holger Czukay, Irmin Schmidt**)

SUICIDE

Born in the pre-punk cauldron of mid-70s NYC, Suicide - **Alan Vega** and **Martin Rev** - was the synthesized equivalent of **the Ramones**, stark minimalism cut to its lowest possible pulse. Introduced via a frantic "Rocket USA" on the 1976 Max's Kansas City album, an eponymous album in 1978 set the critical establishment raving, but when Suicide toured Europe with **the Clash** and **Elvis Costello**, their very presence caused riots - punks did not like synths!

Shaken but not perturbed, Suicide allowed one of their less disturbed shows in Belgium to be released as 24 Minutes over Brussels; a slightly later performance is included on the Half Live cassette - the other half comprises studio out-takes. Ghost Riders captures the duo's 1981 10th Anniversary show.

In 1980, Rev and Vega released a second Suicide album, this time credited to themselves; produced by **Cars** mainstay **Ric Ocasek**, it owes little to Suicide's original power and vision, a failing which the subsequent A Way of Life and Why Be Blue albums also share.

That said, Suicide have remained a major, positive influence on the electronic scene, with acts as disparate as **Soft Cell**, **Foetus** and the **Sisters of Mercy** (whose splinter Gift album Vega appears on) crediting them for their original groundbreaking work. Indeed, at one time so many UK-based artists were citing Suicide as inspiration that it seems difficult to believe that their first shows produced such loathing. Hmm, funny thing, retrospect. **(see Goth, Soft Cell, Synthipop)**

Discography: Suicide (Red Star 1978); 24 Minutes Over Brussels (Bronze 1978); Suicide (ZE 1980); Half Alive (ROIR 1981); Ghost Riders (ROIR 1986); A Way of Life (Chapter 22 1988); Why Be Blue (Brake Out 1992).

DONNA SUMMER

Linking with producer **Giorgio Moroder**, the homophobic queen of disco recorded "I Feel Love" in 1977. The first truly computerized/synthesized dance record, "I Feel Love" was a major international hit, paving the way for countless future imitators/ innovators. Summer herself was unimpressed by her record's impact, and failed to create any kind of worthwhile follow-up. Moroder, on the other hand, has spent much of his subsequent career (with everyone from **Sparks** to **Human League** trying to do as much.) **(see Human League, Kraftwerk, Gary Numan, Punk Rock, Synthipop)**

Swans

SWANS

"If I wanted to hear Black Sabbath at half speed, I'd play Vanllla Fudge. If I wanted to hear Vanilla Fudge at halfspeed I'd play Cop".
– Dave Thomas, Melody Maker 1984

Punks in apocalyptic clothing, Swans' earliest musical output - a more or less constant stream of grlnding hardcore played extremely slowly, extremely seriously, and extremely extreme - is less influential than the theory behind it: the reduction of rock'n'roll/ hardcore to its lowest, loudest, common denominator. After more than a decade of solid noise, Swans remain one of the names to drop in industrial circles, in much the same way as the Stooges and Velvet Underground were during the punk era.

The New Yorkers' self-released Swans EP set the stage for the barrage to come; subsequent releases, leading up to the live Feel Good Now are notable more for their continuity of sound than any development within the band's vision, a phase which abruptly ended when Swans signed to MCA in 1989 and recorded the slick The Burning World with Bill Laswell. ~A year previous, the Love Will Tear Us Apart EP found Swans joining the lengthening list of artists who've attempted to make sense of Joy Division's much overrated greatest hit. "We wore that sound out," vocalist Michael Gira confessed. "I was also seeing that we were attracting a heavy metal crossover audience (and) they can kiss my ass." Since then, Swans have remained on a remarkably tuneful course, almost accidentally developing into a fairly useful pop group.

Selected discography (the noisiest years): Swans EP (1982); Filth (Neutral 1983); Cop (K.422 1984); Swans EP (Homestead 1985); Greed (PVC 1986); Holy Money (PVC 1986); Children Of God (Caroline 1988); Feel Good Now (Product 1988).

SWEET EXORCIST see Richard Kirk

SYNTHIPOP

Duran Duran

A now derogatory term first used to describe the burgeoning New Romantic scene which swept Britain during the early 1980s, pioneered (and best exploited by) the likes of **Soft Cell**, **Depeche Mode** and **Duran Duran**, but also appropriate to the more dynamic work of one time **Residents**' labelmate **Yello**, and Japan's **Yellow Magic Audience**.

The genre's lineage, whilst directly linked to the German bands who broke through in the UK a decade previous (and the ubiquitous "I Feel Love"), was perhaps most heavily influenced by the lowering prices of synthesizer equipment as the 1970s progressed. Hitherto a mysterious noise maker which only **Roxy Music**'s **Brian Eno** had truly learned to manipulate, by the end of the decade there were genuine fears amongst Britain's Musicians Union membership that the synthesizer would eventually take over from the musician completely. The "Keep Music Live" campaign, engineered though it was by hoary old pub rockers who knew three chords and a bottle were already an endangered species, gathered considerable momentum; whether it stemmed, or hastened, the tide of earnest young men with knob-twiddling fingers remains an unanswerable question.

In the post-punk era, with the savage jolt of **Pistols**-whipped adrenalin having already consigned earlier synth experiments (**ELP**, **Barclay James Harvest**, **Yes**, **Uriah Heep** et al) to the scrapyard, the trailblazers were **Ultravox**, then led by the impossibly enigmatic **John Foxx**, **Japan** and **Rikki & The Last Days of Earth**, the only one of the three who truly threatened to take the music into fresh pastures, albeit ones decimated by a particularly unwholesome nuclear winter. Led by **Rikki Sylvan**, the Last Days were pure post-apocalyptic noise masquerading as pop, and which, spread over both sides of their 4 Minute Warning album (DJM 1978), succeeded in alienating both sides of the punk versus pomp debate. Today best regarded as a kitsch classic, the album (and two accompanying 45s) was to remain a red light to other, similarly inclined, groups: no-one dared walk where the Last Days had been.

Three years later, Sylvan was back with The Silent Hours (Kaleidoscope 1981), a set which proved that while he had learned from his mistakes, the rest of the world had still to forgive him for them. Instead, those elements of 4 Minute Warning which had been deemed palatable (the driving thrash of the instrumental "For The Last Days", the sinister smoothness of "No Wave") had now been incorporated into both Ultravox and Japan's musical arsenal, albeit with something less than permanency: while Ultravox drifted into genuine Emerson Lake and Pompous territory, **David Sylvian**'s **Japan** developed a post-**Bowie**/pre-**Current 93** fascination with Far Eastern textures, and ended up constructing mood music for cross-cultural dissidents.

Classix Nouveau

Suicide followed Sylvan into disgrace, albeit of brief duration; by the time the band, tired of bashing its head against a wall of apathy, disbanded, their debut album Suicide was being widely feted for the masterpiece it was.

OMD

But it was **Gary Numan** who finally broke the scene open, in 1979, albeit in a grotesquely manufactured manner; joined by a post-menopausal **Human League**, he laid bare a social underbelly which even its own inhabitants had scarcely been aware of. Striking back at the cultural austerity of punk's savage legacy, a network of clubs sprung up, often in the most unlikely of places, upon which descended an audience which on first inspection appeared to be comprised exclusively of Bowie clones (they even had their own house group, led by one **Ziggy Heroes**).

United by a common love for electronic music, the clubs' deejays - an equally exclusive crowd, amongst whose ranks walked future Some Bizzare label chief **Stevo**, and several members of the New Romantic supergroup **Visage** - propigated a unilateral banner of "Futurism". The nascent **Spandau Ballet**, Liverpudlians' **Orchestral Maneuvers** and **Dalek I Love You, Flock of Seagulls** and **Zaine Griff** all passed through the arena, some already armed with synthesizers, others espousing them altogether, but all united under the same flag. It was the tireless efforts of this circuit which broke **Kraftwerk** into the UK chart marketplace during 1981.

The other principle link was visual - like the Mod scene which had flourished amongst much the same age group two decades earlier, it was a furious, cutthroat business. To paraphrase that earlier era's greatest chronicler, Nik Cohn, if you were seen in last night's haircut you were dead. This was a lesson even **Ministry**, not quite the bad boys of rock they turned out to be, paid attention to!

But some great music did emerge from the foppery, although Duran Duran's "Planet Earth" was as bizarre a precursor of the adventurously percussive "Wild Boys" as Spandau's Kraftwerkian "Muscle Bound" was to "Highly Strung". At the more political end of the spectrum, **Heaven 17** (formed from the divorced ranks of the pre-Dare Human League) cut the immensely prophetic "Crushed By The Wheels of Industry", heralding in intent, if not delivery, the arrival on the scene of a fresh wave of German bands, who really did know how to crush things with industry.

Ultravox

That only Depeche Mode, and to a lesser extent Soft Cell, were to survive the growing weight of **Einsturzende Neubauten** and the like was both a delightful surprise and monotonously predictable - the former because Depeche had hitherto been cranking out single after single of inoffensive boy-meets-girl Eurotat; the latter because in record company head **Daniel Miller,** they possessed a mentor whose own track record included two of the harshest "songs" ever flung into the singles' market - "TVOD" and "Warm Leatherette" (later covered, incidentally, by disco amazon **Grace Jones**).

Today the impact, and - were one to go back in time to the days when the New Romantics were in full swing - the sheer horror of much of the music perpetrated in its name, has been largely forgotten; either that or absorbed wholly into the body corporate of rock'n'roll history. Even the absolute abberations which marked **Robert Palmer** and **Alice Cooper**'s entries into the field have today been accepted as such, while the challenging abrasiveness of **PiL**'s Metal Box was aimed at least in part at the more obtuse end of the market.

Even at its most derivative (and the likes of **Howard Jones** and Flock of Seagulls were never going to be anything but), then, New Romanticism was a rare example of a musical movement which was wholly self-sustaining. Even the purging power of punk had drawn on the popular past for both inspiration and material (the Pistols cut their teeth on old Monkees and Who songs, **Sham 69** recorded **Yardbirds**' b-sides, and everybody loved the **Velvets** and the **Stooges**); the New Romantics literally started with a clean slate, with even their most obvious points of reference - Kraftwerk, Berlin-era Bowie, Eno, etc - open-ended enough to allow at least a hint of originality to appear. With the example of Rikki Sylvan paramount in the mind, it is just a shame that much of the best music remained unheard. **(see Depeche Mode, Einsturzende Neubauten, John Foxx, Human League, Ministry, Gary Numan, Kratwerk, Soft Cell, Suicide, Donna Summer)**

TACKHEAD

Tackhead

New Jersey based Tackhead was formed by members of the Sugar Hill rap label's house band - **Skip McDonald**, **Keith LeBlanc** and **Doug Wimbish**, plus producer **Adrian Sherwood**. Early single-only releases appeared under the name **Fat Comet**; the quartet also worked extensively with former **Pop Group** maestro **Mark Stewart**.

The first Tackhead album, Major Malfunction, was released under LeBlanc's name in 1986, a belated follow-up to his "No Sell Out" dance 12" of three years previous. The following year, the group joined with British MC **Gary Clail**, under the name **Gary Clail's Tackhead Sound System** and released the live Tackhead Tape Time, a convoluted collage of samples, shouting and scratchy guitars. A second Clail album, End of the Century Party, also utilizes Tackhead members.

1989's Stranger Than Fiction, a second LeBlanc album, is Tackhead by default - all the members turn up somewhere. But the first "proper" Tackhead release, with vocalist **Bernard Fowler**, was the same year's Friendly As A Hand Grenade; it was also the last to capture Tackhead as the energetic experimentalists who had hitherto flourished. Subsequent releases from the Tackhead stable are industrial bark, with no substantive bite.

Select discography: Major Malfunction (**Keith Leblanc**: World 1986); Tackhead Tape Time (**Gary Clail** etc; Nettwerk 1987); End of the Century Party (**Gary Clail** & On-U Sound; On-U 1989); Stranger Than Fiction (**Keith Leblanc**; Nettwerk 1989); Friendly As A Hand Grenade (TVT 1989); Strange Things (SBK 1990)

TANGERINE DREAM

Formed in 1967, but only gradually taking shape over the next two or three years, Tangerine Dream differed from many of their German contemporaries in that their earliest work was most heavily influenced by **Edgar Froese**'s personal love of **Jefferson Airplane**, **the Doors** and **Jimi Hendrix**, together with the inevitable **Floyd**. However, he also

Tangerine Dream

enjoyed avant-garde jazz, and the first Dream album, Electronic Meditation is a collision of the two abetted by sundry special effects - burning parchment, a whip and broken glass included.

The band at this stage were still employing conventional instruments; the shift towards pure electronics was gradual, and the second album, 1971's Alpha Centauri is most notable for its pioneering use of echo as an instrument rather than an effect. The side long title track, however, is impressive, as organ and flute fight in slow motion over a sharp electronic backdrop.

With drones and tones gradually taking precedence over melody, Zeit and Atem powered the Tangs further into the realms of musical purity than any other band had dared go; by the time of Phaedra, the album which broke the group on a truly international level, all three musicians - Froese, **Chris Franke** and **Peter Baumann** - were completely synthesized, and describing their music as an absolutely spontaneous creation, bound only by time constraints. Thus Phaedra and Rubycon become 40-odd minutes of sound, utterly unique to the moment; likewise the band's concerts - held, amidst much publicity and controversy - in sundry European cathedrals. One did not attend a Tangerine Dream concert and call for one's favorite songs.

Ricochet, drawn from these concerts, remains the pinnacle of the band's aspirations; beginning with Stratosphere, and continuing to this very day, Tangerine Dream have become somewhat more regimented, to the point where both technology and the musicians' own limitations, have overtaken them. Nevertheless, both Tangerine Dream and, as an occasional soloist Edgar Froese, will always remain an integral, and important, component in the development of the industrial field. (**see Kraut Rock**)

Selected discography: Electronic Meditation (Ohr 1970); Alpha Centauri (Ohr 1971); Zeit (Ohr 1972); Atem (Ohr 1973); Phaedra (Virgin 1974); Rubycon (Virgin 1975); Ricochet (Virgin 1975).

Test Department

TEST DEPARTMENT

Formed in the first flood of enthusiastic mimicry which followed **Einstrzende Neubauten**'s early UK breakthrough, Test Department made similar journeys through pure industrial sounds on their first album (Beating the Retreat), before allowing such disparate sounds as cello, harp and a 90 strong Welsh choir to move into the landscape on Shoulder to Shoulder, an album which highlights the depth of feeling which surrounded the crippling coal miners' strike then paralyzing Britain. **Robert Calvert**'s Freq is similarly conscious of this unprecedented shattering of the social order, and his use of recorded interviews with miners was echoed by Test Department on The Unacceptable Face of Freedom, a powerful if somewhat conventional album - espousing the hammering and drilling of old, Test Department now used synthesizers, sequencers, drums and - surprisingly - bagpipes to make much the same point. Much of this album resurfaced in live form on the retrospective Materia Prima concert recordings.

Recorded live in London and Amsterdam, A Good Night Out offered Test Department a brief respite, as well as a welcome opportunity to place their past activities into some sort of coherent framework. Terra Firma completed this process, a radical political tract set to music.

Gododdin, a collaboration with Welsh theater group **Brith Gof**, is based upon an early Welsh epic poem, and was staged in a disused car factory in Cardiff.

Discography: Beating the Retreat (Some Bizzare 1984); Shoulder to Shoulder (with the **South Wales Striking Miners Choir** Some Bizzare 1985); The Unacceptable Face of Freedom (Some Bizzare 1986); A Good Night Out (Some Bizzare 1987); Terra Firma (Play It Again Sam 1988); Gododdin (with **Brith Gof** Ministry of Power 1989); Materia Prima (Dept 1 1990).

JIM THIRLWELL see Foetus

THIS HEAT see Punk Rock

THROBBING GRISTLE

"The basic delusion is that we're a music group. That's where they start to go wrong" - Genesis P-Orridge, 1978

Throbbing Gristle

Six years after they met in the northern English city of Hull, performance artists **Genesis P-Orridge** and **Cosey Fan Tutti** formed Throbbing Gristle in September, 1975, performing a musical hybrid which lay half in the still embryonic punk rock camp and half in traditional performance art. A minute and a half of the band's first ever live show was included on an Italian various artists album, Mission Is Terminated, together with two TG tracks of later live vintage ("You Don't Know"/"Damura Sunrise", packaged on a bonus 12" single), and two further cuts remixed for Italian radio (other tracks from this source appeared on Journey Through A Body in 1982).

It was to be a year before they first received any attention, when P-Orridge and Tutti (under the name **Coum Transmissions**) staged a retrospective exhibition, Prostitute, at London's ICA in October, 1976 (support for the show was provided by punk band **Chelsea**, making their first ever live performance). In the ensuing controversy, the couple's Art Council grant was terminated, and they were banned from exhibiting at any UK galleries.

The group's self-confessed interests included torture, cults, wars, venereology, concentration camp behavior, Aleister Crowley, unusual murders, and unusual pornography. It was not necessarily a sick collection, however, rather, in latching onto some of society's greatest tabboos, the themes which polite circles never touch upon, the group was arguably opening up areas in which understanding did not have to go hand in hand with outrage. "We're just troublemakers really, cause otherwise the world's a very boring place to be."

Throughout the first year of the band's existence, almost every note they played, at jams and rehearsals alike, seems to have been recorded. New TG volumes one and two, Nothing

Short of a Total War, Time To Tell and 23 Drifts to Guestling, all posthumous releases, document these important formative months. The last named is of additional interest in that it includes one track from Genesis P-Orridge's first ever recording, a 1969 album called The Early Worm. Just one copy of the record was pressed (it cost him eight pounds); a projected follow-up, The Early Bird Catches was never completed.

Turning their attention instead to film, the Coum Transmissions project was continued with 1977's After Cease To Exist, a ten minute film for which Throbbing Gristle - P-Orridge (violin, vocals, bass), Tutti (guitar, cornet, effects), **Peter Christopherson** (tapes, trumpet) and **Chris Carter** (synthesizer, rhythms) - produced a twenty minute soundtrack, released over one side of their second album, 1977's 2nd Annual Report. Their debut, The Best Of... Volume Two was a limited edition of thirty cassettes, circulated amongst friends only. It was later reissued, and three years later spawned the single "We Hate You" (Sordide Sentimentale)

2nd Annual Report, too, was a limited edition (785 copies, with a larger, 2,000 strong, reissue following on Fetish in December 1978), and aside from the soundtrack, it also featured four early live cuts and two studio takes. Record Collector magazine later described this side of the album as amongst "the most exhilarating 20 minutes of music ever committed to vinyl". A new edition of this album was released following the band's demise, recut with the music playing backwards (Fetish). A chamber music ensemble was also added to side two, creating a grotesque, but somehow very effective, collage of sound. This edition also appeared, alongside three subsequent Gristle albums and a live set, Mission of Dead Souls, in A Boxed Set, a five disc collection released in October, 1981.

"United", Throbbing Gristle's first single, broke the band into the UK independent charts; it also appears dramatically recalibrated (it has been sped up until it lasts just 16 seconds) on DOA: The Third and Final Report, the third Throbbing Gristle album.

Released on the group's own Industrial label, this latest Report again combined live and studio work, although not without some surprises. One of the live tracks, "Wall of Sound", was pieced together from five separate performances, while four of the studio pieces were solo projects by the individual band members -shades of **Pink Floyd**'s Ummagumma! Another cut, "Death Threats", comprised of answer-phone messages left by a writer for the UK magazine Zig Zag and Chris Carter's wife, who believed that Tutti (who had by now split up with P-Orridge) was leading her husband astray. Her suspicions, incidentally, were later proven correct.

After three years spent as a potent, challenging and often difficult underground force, 1979 saw Throbbing Gristle make what were apparently very genuine attempts to break out of the sick cult mode into which many people continued to cast them. Twenty Jazz Funk Greats was their most accessible album yet (although there were, in fact, only two songs which could even remotely be called jazz-funk, opening each side of the record), and a series of gigs saw the band trouping out in smiling white - a far cry from the scowling militaristic black of the past. (A souvenir of one of the earlier shows, at the Crypt in November, 1978, was released by Mental Decay as Special Treatment in 1984.) Their reputation was seemingly assured when **David Bowie** announced on US radio that Throbbing Gristle was the most important thing then happening in the UK.

At the same time, the band was still aware of its heritage, as the release, by

Industrial, of the cassette only Pastimes/ Industrial Muzak collection of unreleased material makes clear. There was also a boxed set of similar material, the 24 Hours collection of 24 hour long cassettes, and in 1982, future Psychic TV member Geoff Rushton released Assume Power Focus, which brought to light more unreleased studio material. This particular aspect of the group's career was completed with the CD release of TG CD ! in 1986, a collection of unreleased 1979 recordings.

Despite their growing reputation as a live act, Throbbing Gristle's first official live album, 1980's Heathen Earth was recorded in the studio, in front of a group of invited friends. The performance was also caught on video, and released through Industrial. (It has since been reissued, in censored form, alongside Live at Oundle Public School). A second live album, Psychic Rally in Heaven (Rough Trade) was recorded at the gay night club, Heaven, in January 1981. Later, following the band's break-up, this same period was documented by a slew of in-concert recordings, including Music From The Death Factory May 1979 (Death), The Psychic Sacrifice (Karnage) and its abridged companion Sacrifice (Castle), Funeral In Berlin (Zensor), Fuhrer der Mensheir (American Phonograph), Once Upon A Time (Casual Abandon) and Editions Frankfurt - Berlin (Svensk Illuminated). Several of these releases, incidentally, have been tagged as bootleg recordings by some discographers.

The changing musica face(s) of Throbbing Gristle was highlighted on the two double A-sided singles released simultaneously in September, 1990. Techno-pop ("Adrenalin"), lilting synth music ("Distant Dreams"), reasonably straightforward rock ("Something Came Over Me") and uncompromising industrial roar ("Subhuman) all raised their heads, with a third single, the live "Discipline", blending all four styles into one lengthy, compulsive whole. (A studio version of this track, with vocals by **Soft Cell**'s **Marc Almond**, was released by the Flexipop magazine.)

In May, two months after P-Orridge married his new girlfriend, Paula, Throbbing Gristle split up, bowing out with two shows in California (the last of these shows comprised the Mission of Dead Souls live album included in the A Boxed Set release). A compilation, Throbbing Gristle's Greatest Hits - Entertainment Through Pain (Rough Trade) was released in the US a little later

"Assuming that we had no basic interest in making records, no basic interest in making music per se, it's pretty weird to think that we'e released something like ten albums plus bootlegs; 40 cassettes...that have had an effect on the whole popular music scene, forever...." - Genesis P-Orridge 1982

Carter and Cosey immediately formed their own band, the **Creative Technology Institute**; P-Orridge, Christopherson and Paula remained together as **Psychic TV**. **(see Chris & Cosey, Coil, Psychic TV, Punk Rock, Boyd Rice, Z'ev)**

Selected discography: The Early Worm (P-Orridge private 1969); The Best Of... Volume Two (Industrial 1977); 2nd Annual Report (Industrial 1978); DOA: The Third and Final Report (Industrial 1979):; Twenty Jazz Funk Greats (Industrial 1979): Heathen Earth (Industrial 1980): Throbbing Gristle's Greatest Hits - Entertainment Through Pain (Rough Trade 1980); Mission of Dead Souls (Fetish 1981); A Boxed Set (Fetish 1981); Journey Through A Body (Walter Ubricht 1982); Psychic Rally in Heaven (Rough Trade 1982); Assume Power Focus (Power Focus 1982); Mission Is Terminated (Expanded 1983); Special Treatment (Mental Decay 1984); TGCD 1 (Mute 1986).

TILLER BOYS

Buzzcock Pete Shelley, a self-confessed **Can** fanatic (he penned the sleeve notes to a 1981 compilation), linked with two friends **Francis Cookson** and **Eric Random** to recreate **Fripp/ Eno**'s No Pussyfooting, both in concert and on a three track EP. Generally scorned for its improvisational esotericism, the project utilized pre-recorded tapes, drum machines and random electronics to create a collage of directed noise, much of which could be generated with no human particpation whatsoever. It was reported at the time that elements of the work dated back to 1974.

Random's solo career, confined to a handful of singles ("Dow Chemical Company", "23 Shidoo" and "That's What I Like About Me") and two albums (released, like the Tiller Boys' effort, on Buzzcocks' own New Hormones label), depended on similar tolerance, with the performer's own adopted surname perhaps the best indication of the success of his work. Random was later involved in **Cabaret Voltaire**'s **Pressure Company** project.

Shelley later pursued similar, but more tuneful directions with his post-Buzzcocks career. The single "Homo-Sapien", from the album of the same name, is the best of his solo prolusions, a raucous controlled noise over which Shelley chants his lyrical mantra. (**see Cabaret Voltaire, Can, Punk Rock**)

Discography: TILLER BOYS Big Noise from the Jungle EP (New Hormones 1980); Sky Yen (**Pete Shelley** Groovy 1980); Homo-Sapien (**Pete Shelley** Genetics 1981).
ERIC RANDOM That's What I Like About Me EP (New Hormones 1980); Earthbound (New Hormones 1981); Live (New Hormones 1982)

FRANK TOVEY

The "Back to Nature" single was the second-ever Mute Records release (following **The Normal**), and established Tovey, then operating under the **Fad Gadget** alias, as a genuinely uncompromising electronics whizz, whose eye for musical effect never disturbed his talent for witty arrangements and melody.

"Ricky's Hand" followed, a slice of heavy atmosphere on which **Daniel Miller** contributed further synth; Fireside Favorites, an album which made a very successful grab for the **Human League**'s recently abdicated throne, and the darker Incontinent are both recommended. Completists will also want the "Fireside Favourites"/"Insectide" and "Make Room"/"Lady Shave" 45s.

Sadly the final Fad Gadget albums did not match their predecessors, either in content or style (the presence of soulful songstress **Alison Moyet** on the former was particularly galling), and as if in recognition of this, Tovey retired the Fad project and concentrated upon his "own" career. (The Singles collection is an enjoyable, if offbeat, look back at both Tovey and Mute's darker side.)

Aside from a live album recorded with **Boyd Rice**, Tovey's solo career distanced itself considerably from Fad Gadget, utilizing many of the same tricks as **Wire** and early **XTC** were prone to, albeit strictly on Tovey's own terms. Even at his worst (Civilian), he remains innovative; Tyranny is particularly strong, translating American folk ballads into electronic language with devastating results. (**see Daniel Miller, Boyd Rice**)

Discography: Fireside Favourites (**Fad Gadget** Mute 1980); Incontinent (**Fad Gadget** Mute 1981); Under The Flag (Fad Gadget Mute 1982); Gang (**Fad Gadget** Mute 1984); The Fad Gadget Singles (Mute 1986); Snakes & Ladders (Mute 1986); Civilian (Mute 1988); Tyranny & The Hired Hand (Mute 1990); Grand Union (Mute 1991).

TREPONEM PAL

Vocalist **Marco Neves** describes the French band as "Industrial Metal", an apt description for a group brought together through their friendship with **Young Gods**. Formed in Paris in 1986, and taking their name from a syphillus virus, Pal's first three song demo (produced by God **Franz Treichler**) found its way to a journalist friend, who in turn passed it on to Roadrunner Records. Treichler subsequently produced Treponem Pal's first, self-titled, album.

Treponem Pal's bonecrunching clatter only increased on Aggravation, an album which is also remarkable for its disemboweled cover of **Kraftwerk**'s "Radioactivity". (**see Young Gods**)

Discography: Treponem Pal (Roadrunner 1989); Aggravation (Roadrunner 1991).

TUBEWAY ARMY see Gary Numan

TUBALCAIN

In the winter of 1991/1992, former **Executive Slacks** members **John Young, Robb Jordan** and **Athan Maroulis** began forming Tubalcain. The name Tubalcain, was culled from the members ongtime interest in masonic ceremonies and symbols. Tubalcain is credited as being the first mason in the bible. Although similar in it's tribal, often primal approach, Tubalcain screams above the **Executive Slacks** minimalist style. The often copied line-up of machines and musicians, originated by **Executive Slacks**, can be seen and heard here. This line-up includes the use of samples, drum machines, stand-up drum kit, guitar, vocals and the newly added live bass.

Tubalcain

After a stint of live shows in Washington D.C., New York City and Philidelphia, Tubalcain flew to Los Angeles to record their first album, 25 Assorted Needles. The first release from that album was a limited edition 12", featuring a cover of "2000 Light Years from Home". In the winter of 1992, Tubalcain released their first full length album, 25 Assorted Needles, on compact disc and cassette. As I write these words, it is recieving a considerable amount of press and airplay in both Europe and America. After a series of live dates with **KMFDM** and **Chem Lab**, the band plans on residing in Los Angeles and working on their next album. (See **Spahn Ranch**).

Discography: 2000 Light Years from Home (Verdugo Records 1992); 25 Assorted Needles (Verdugo Records 1992).

TUNNELMEN

Formed in Swindon, England in 1988 by **Biff, Jeff** and **Audrey,** Tunnelmen played gigs and demoed the U.K. in early 1990 the band visited the USA and gigged in Los Angeles for six months, building a strong following on the underground scene before returning to

Tunnelmen

England due to visa limitations. The band recorded their debut album for an independent L.A. label, Nate Starkman & Son. The album Defeat The Inefficiency Of Crashing received critical acclaim but little distribution and no publicity. The label and distributors parted company and the album was deleted.

Select discography: Defeat The Inefficiency Of Crashing (1990 Freak).

NIK TURNER'S SPHYNX see Robert Calvert, Chrome, Punk Rock,

ULTRAVOX see John Foxx, Synthipop

VAMPIRE RODENTS

We really have no great love or understanding of Industrial music. We picked the genre for its characteristics" - **Anton Rathausen, 1992.**

Sinister sonic terrorists from Phoenix, Arizona, **Anton Rathausen** and **Victor Wulf** met in Toronto during the early 1980s, and in 1988 decided to combine Rathausen's penchant for film music with Wulf's experimental soundscapes under the name Vampire Rodents.

Describing their music as "sample-based composition" ("but that's seven syllables and you're only allowed four"), the Rodents are reminiscent of **the Residents** in their often militaristic harshness - "Babelchop" and "Babyface", the opening cuts from Premonition; but elsewhere delve into less confrontational moods, to gripping effect. Further comparisons, to **Coil**, **Test Department**, **Wire** and **Skinny Puppy**, are relevant, but again demand qualifications which only a good earful can provide. A forthcoming collaboration with **Babyland**, **Recline** (on the Rivet Head compilation) sounds promising. **(see Babyland)**

Discography: War Music (VR, 1990); Premonition (VR 1992); Lullaby Land (Re-Constriction 1993).

VAN DER GRAAF GENERATOR

The project around which maverick chest-beater **Peter Hammill** based much of his less introspective material, VDGG were very much part of the progressive rock movement of the early-mid 1970s. Their avant-garde credentials were established around only a handful of pieces, "Pioneers Over C", from 1970's H to He album; and the side long "Plague of Lighthouse Keepers", from the preceding The Least We Can Do, being the most noticeable. Amongst the many effects employed in the latter piece, the sound of a ship crashing into rocks is certainly amongst the most haunting in the entire industrial field.

Frustratingly inconsistent, much of VDGG's work today sounds indulgent and pretentious, which of course it usually was. However, alongside fragments of Hammill's own work, most notably The Future Now/PH7 coupling ("Motorbike in Afrika" from the first named is of particular sonic relevance), they remain massively influential, with **John Lydon** amongst the most outspoken of their proponents. (**see Public Image, Punk Rock**)

Selected discography: VDGG The Least We Can Is Wave To Each Other (Charisma 1969); H to He Who Am The Only One (Charisma 1970); Pawn Hearts (Charisma 1971)
PETER HAMMILL: The Silent Corner & The Empty Stage (Charisma 1974); The Future Now (Charisma 1977); PH7 (Charisma 1978).

EDGARD VARESE

Born in 1883, Frenchman Varese anticipated many of the concepts of electronic music. Substituting sound complexes or traditional harmony, his early compositions are notable for the intensity and volume of the percussive effects.

Ecuatorial, in 1934, was Varese's first electronic composition of note. He died in 1965.

Selected discography: Hyperprism (1923); Octandre (1923); Integrales (1925); Arcane (1927); Ionisation (1931); Ecuatorial (1934); Poeme Electronique (1958).

VEIN CAGE see OUT OUT

VELVET UNDERGROUND

Velvet Underground

Lou Reed and **John Cale**'s own backgrounds could not have been more disimilar; while Reed worked as a wage-slave songwriter for Pickwick Records, Cale was studying avant garde classical music and working with **LaMonte Young**. The meeting of these two minds, then, was not one many people would have considered potentially productive; however, under the aegis of artist **Andy Warhol**, and with the band's ranks swollen by **Maureen Tucker** (drums), **Sterling Morrison** (guitar) and briefly, **Nico** (vocals and harmonium), the Velvets launched on a sonic journey whose true ramifications have still to be fully understood.

The first two albums are the most important: Andy Warhol's... boasts "Black Angel's Death Song" and "European Son to Delmore Schwarz", improvisational monsters which took rock'n'roll into the realms of total aural abuse. The corollary to these pieces, of course, were the straightforward, and infinitely better known "Waiting for the Man" and "Sunday Morning". "Venus in Furs" provides a comfortable middle ground.

White Light White Heat, recorded without Nico, is even heavier - Lou Reed subsequently claims to have invented heavy metal with the 17 minute "Sister Ray"; he might lay similar claim to industrial music, the hammering, yammering monotony of "Ray" remaining a non-musical freak-out milestone for at least the next decade. Elsewhere, the guitar solo in "I Heard Her Call My Name" is the perfect echo of the preceding lyric - "and then my mind split open".

The final Velvets' albums, whilst equally influential, see Reed taking his chances from within conventional song structures ("Murder Mystery" from the third album, for instance), and the band's break-up came hot on the heels of their most commercial album yet, Loaded. Arguably, had the band maintained that particular direction to its logical pop star conclusion, the legend of the Velvets might well have been stillborn.

In the event, the frenzied name-dropping which **David Bowie** indulged in during his first flurry of post-Ziggy stardom ensured that the Velvets would never lie still; Reed's and Cale's subsequent high profiles were similarly to keep the band's name alive, and three times over the next decade, in the darker gleamings of the Glam movement; five years later in the white heat of punk; and lastly, at the experimental dog-end of the New Romantic movement, the Velvet Underground were touted as masters of the new art form.

Beyond their conventional recordings, the Velvets have been very well-served by bootleg and semi-official compilations; quality varies, but a safe rule of thumb for the avid noise-seeker is the earlier (1965-67) the better. Live 1969 and the Another View/VU collections are recommended only to fans of Reed's more melodic moods. **(see also David Bowie, John Cale, Metal Machine Music, Punk Rock, LaMonte Young)**

Selected discography: Andy Warhol's Velvet Underground (Verve 1966); White Light White Heat (Verve 1967).

VOICE OF DESTRUCTION

Sonic terrorists based in San Francisco, Voice of Destruction is the brainchild of **Lewis Cypher**, ex-of industrial primitives **Jon Jon**, and former **Premonition** vocalist **Clawed**. Joined by **Cousin Enos**, VoD debuted with Steamroller Tactics, an aptly named album described in Alternative Press as "a Gothic **Motorhead** as visualized by **Ministry**." Already threatening to become a major force on the industrial scene, (**see Goth, Jon Jon**)

Discography: Steamroller Tactics for Fun and Profit (Cleopatra 1992); Black Are The Souls Of The Damned (Cleopatra 1993).

WARSAW see Joy Division

WARWORLD

Seattle's **Mikhail Bohonus** is the mastermind of Warworld, a studio-bound concept for creating "soundtracks for nightmares." Tracks are characterized by churning, syncopated rhythm patterns and skanky guitar damage with distorted samples stabbing through it all. Much of Bohonus's later material is primarily instrumental, and indeed has an eerie 'film score' quality to it.

Warworld has earned a respectable underground reputation from the release of the 1991 RRR vinyl release of the same name, as well as for his numerous cassette releases.

Selected discography: Spots On My Wall (Warworld 1988); Trench Mouth (Warworld 1989); Interstate (Warworld 1990); Warworld (RRR 1990).

ROZZ WILLIAMS see Goth, Premature Ejaculation

WIRE

"One is not supposed to enjoy a Wire gig." - **Colin Newman**, 1979

Although they are widely regarded as part of the second wave of British punk acts, Wire constantly distinguished themselves in the days of xerox and spit with a harsh, robotic sound which seemed completely at odds with all else that was going on. Certainly their appearance, barely rehearsed, and virtually unknown, on the seminal Live At The Roxy WC2 album was anomalous, their versions of "12XU" and "Lowdown" a vicious contrast to "Bored Teenagers" and "Oh Bondage! Up Yours!", two of the other songs on the record.

Picking up pace in the wake of the album's release, Wire - Colin Newman (vocals, synthesizer, guitar), **BC Gilbert** (guitar), **Graham Lewis** (bass) and **Robert Gotobed** (drums) - continued their policy of musical alienation, prompting one newspaper to describe their "mekanical vokalist" as being "so b-l-a-n-k he was a vacuum". When Wire opened for American glam rockers **the Tubes**, the concert hall's bar had not been so busy in months.

The band's first demos, most of which were to be posthumously released as Wire Play Pop (Pink Label) were recorded during the summer, and in September, 1977, Wire signed to Harvest, and released their first single, a three track effort led off by the ferocious "Mannequin" in November. All three songs resuraced on Pink Flag, the band's awesome 21 track debut, later that same month.

"I Am The Fly", still Wire's most crucial recording, followed in the new year, with the similarly frenetic "Dot Dash" arriving shortly before the band's second album, Chair's Missing.

A more reflective col ection (it boasted only 15 cuts, but neither of the last two singles), Chair's Missing provided Wire with their first and only hit single, "Outdoor Miner". Caught, however, amidst accusations of chart hyping, the single swiftly disappeared from the listings, and Wire returned to the studio to begin work on 154.

154 (which was originally released with a free four track EP) and its attendant single, "Map Ref. 41N 93W" was intended only as Wire's final Harvest release. It also became their last studio release. (CD releases of all three Wire albums include bonus tracks; On Returning compiles the very best of these three albums, and, bar the omission of the Peel Sessions material, can be considered definitive.)

Although the band returned to the studio following the album's release, debuting several new songs on tour and recording a number of others (a single of "Our Swimnmer"/ "Midnight Bahnhof Cafe" was released by Rough Trade in 1981; the same label's 1983 "Crazy About Love" single was recorded for the BBC's John Peel radio show), the individual members were tiring of Wire's continued lack of commercial success, especially when it was paired with the now salivating recommendations of the music press. There was also, according to Gilbert, a need to "follow up the idea [on 154] in solo projects."

In July, 1980, following a show at London's Electric Ballroom, the band broke up. (The show was recorded and appeared a year later as Document and Eyewitness - Rough Trade. A free 12" featured eight further live tracks, including one - from a gig in Montreux - upon which the loudest noise is what Newman calls "the beautiful...sound of 8,000 Germans whistling and shouting 'Get off!'".)

The individual members went their own separate ways, recording a string of releases which only occasionally follow the tracks laid down by Wire; a 1986 reunion for the Snakedrill EP was somewhat unfulfilling, prompted by the success of the And Here It Is Again compilation, by early 1987 the band was recording its first new album in eight years, the compelling, if a little laid-back, The Ideal Copy.

An EP, Ahead, served up remixes of the title track, plus three tracks recorded live in Berlin; like the live-but-meddled-with It's Beginning To And Back Again album, they showed Wire to have lost little of their early live frenzy, but in the studio both A Bell Is A Cup and the out-takes EP Kidney Bingoes were considerably weaker. Manscape, an improvement in musical terms, also proved Wire had now moved as far from their punk roots as it was possible to get, seeking a more electronic feel.

Drill - named for a fierce five minutes of guitar-shaped scratchings on Snakedrill - continued in a similar vein, often recounting some of its namesake's power. However, it was to be Wire's final release: following drummer Richard Gotobed's departure, the survivors regrouped as **Wir** and recorded the strangely slick The First Letter. **(see also Dome, Bruce Gilbert, Graham Lewis, Colin Newman, Punk Rock)**

Selected discography: Pink Flag (Harvest 1977); Chairs Missing (Harvest 1978); 154 (Harvest 1979); EP (Rough Trade 1980); Document & Eyewitness (Rough TRade 1981); Wire Play Pop (Pink 1986); Snakedrill EP (Mute 1986); And Here It Is Again (Sneaky Pete 1986); The Ideal Copy (Mute 1987); Ahead (Mute 1987); The Peel Sessions (Strange Fruit 1987); A Bell Is A Cup Until It Is Struck (Mute 1988); Kidney Bingoes EP (Mute 1988); It's Beginning To And Back Again (Mute 1989); The Peel Sessions Album (Strange Fruit 1989); On Returning 1977-79 (Restless 1989); Manscape (Mute 1990); Drill (Mute 1991); The First Letter (as **Wir** Mute 1992); The A List (Mute 1993).

WIR see Wire

WISEBLOOD see Foetus

X.M.T.P.

X MARKS THE PEDWALK

X Marks The Pedwalk are convincing with their structured and conceptual soundwork from release to release. Their album Freaks perfectly combines severity and feeling, whereas their 12 inches - like the late Cenotaph - give devotion to danceable industrial collages. X Marks have however topped themselves with Human Desolation. X Marks leader **Sevren Ni Arb** also releases recordings as **Ringtailed Snorter** and **A Head.**

Selected discography: Arbitrary Extortion 12" (Zoth Ommog); Disease Control (Zoth Ommog); Abattior (Zoth Ommog); Freaks (Zoth Ommog); Cenotaph (Zoth Ommog); The Trap (Zoth Ommog); Human Desolation (Cleopatra); Paranoid Illusions (Zoth Ommog).

XON see Richard Kirk

XORCIST

Xorcist, a solo project, was formed by Peter Stone in 1985 when he started to take a more radical and unorthodox approach to writing electronic music. The sound of Xorcist was first forged by an archaic music computer system called the PPG Wave 2.3 synthesizer with a PPG Waveterm–A acting as the sampling and sequencing brain.

Shortly thereafter, Xorcist began using multiple computers of various types along with a barrage of synthesizers, samples, and effects processors to create what Xorcist has coined CyberDustrial music.

Xorcist is a firm believer in that we live in an age where technology has finally begun to geometrically surpass our human learning capacity for understanding the consequences of our technological advances. If we compare human existence against our technological advancement curve, by the year 2000, we will no longer control technology. Technology will control us.

Select discography: So Big (1992 It Moves); Damned Souls (1992 21st Century); Bitches (1993 21st Century).

XTC

Only two XTC albums really matter, the beautifully dissonant and in places aptly titled White Noise, and Go 2/Go +, an album and a half of murky mutant punk which, two-thirds of the way through, turns and disassembles itself in the name of dub. The later poisoned whimsy of **Andy Partridge** and Co. would have been trampled out of existence but for the weight of these two albums.

Also of interest is the debut 3D EP, which included one track ("Science Friction") which would reappear in revised form on White Noise. All three cuts were appended to CD releases of the album.

Partridge has worked with **the Residents** amongst others; fellow founder member **Barry Adamson** went on to **Robert Fripp**'s **League of Gentlemen**. (**see Robert Fripp, Punk Rock, Residents, XTC**)

Select discography: 3D (Virgin 1977): White Noise (Virgin 1978): Go 2/Go+ (Virgin 1978)

XYMOX

The Dutch quartet began life as the heavily derivative **Clan of Xymox**, dripping the era's standard Goth influences - **Bauhaus, Christian Death, Siouxsie**. Their first eponymous release in 1985 was an adventurous disappointment, best compared to **Danse Society** (when it was good) and **New Order** (when it wasn't); the 12" single "Stranger"/"A Day" was considerably better, and set the stage for Medusa, two years later. That album is described by the band as "pretty much the blueprint for everything we've done since."

The Million Things EP debuted the band's abbreviated name in 1988; the following year brought Twist of Shadows, a collection of pure electronic dance numbers. Phoenix is a little harsher, but still shows embarrassing **Telex**-type influences. Undemanding.

Discography: The Clan of Xymox (Clan of Xymox 4AD 1985); Medusa (Clan of Xymox 4AD 1987); Twist of Shadows (Wing 1989); Phoenix (Wing 1991).

LAMONTE YOUNG

American Young was an early avant-garde pioneer whose work included lengthy conversations with a piano, screaming at plants, etc. Early works include For Brass (1958), For Guitar and Voldelta 23 (1959), Raga by Rarra (1959), all collected on I; Poem for Chairs etc and Piano Study (II); Composition 1960 #7 (1961), Bb Dorian Blues and Day of the Antler (1963) (III). Young's early work with **Terry Riley** is collected on V; XI features a performance of Riley's In C recorded in 1989.

Young has seldom recorded with a general audience in mind - his best known album, recorded with the **Dream Syndicate** (allegedly featuring **John Cale**) in 1962, was a limited edition priced in excess of $1,000; another piece, Trio For Strings, has never been recorded, despite its highly publicised influence on the likes of **Steven Reich** and **Philip Glass**. This silence has been broken only recently, with a number of Young works appearing on cassette-only, through the Velvet Underground's appreciation society.

Other musicians to come under Young's direct influence were **Holger Czukay** and **Irmin Schmidt** (of **Can**), and the **Velvet Underground**'s original drummer, **Angus MacLise**. **Lou Reed** also credited Young's experiments in atonality as a direct precursor of Metal Machine Music. (**see John Cale, Can, Holger Czukay, Metal Machine Music, Terry Riley, Irmin Schmidt, Velvet Underground**)

Selected discography: (all released through 50 Skidillion Watts. Dates refer to material performed, not release): I (1958-60); II (1960-61); III (features **John Cale** 1961-63); IV (1961-66); V (features **Terry Riley** 1964-8?); VI (1987); VII (1987); VIII (1987), IX (1987); X (1990); XI (features **Terry Riley** 1989-91).

YOUNG GODS

"Most of our audience comes for the music, not to see someone sweating on the neck of a guitar. There are enough bands doing that already" - **Franz Treichler**, 1992.

Naming themselves after a **Swan's** song, the Swiss electropoppers created a huge stir in 1985 with the release of their first single, "Envoye!". Turning the usual techno format on its head, Young Gods featured a live drummer, and sampled guitar, alongside vocals sung in French by Treichler (the only consistent band member).

Their first eponymous album was unable to match the furor created by the single, although the inclusion of **Gary Glitter**'s "Hello Hello I'm Back Again" at least kept the dance floor shaking, and the British Melody Maker named it their 1985 Album of the Year. However, the release of L'Eau Rouge in 1989 saw the Gods fulfill their earlier promise, a guitar heavy monster which showed someone had been paying close attention to **Ministry**'s latest prolusions.

Young Gods

Play Kurt Weill is a straightforward recounting of the German songsmith's music, although it says little which hadn't already been covered by the multitudinous interpretations of **Weill** on the 1985 Lost In The Stars tribute album (from which the Gods, of course, were absent) missed.

The band's first English-language album, TV Sky saw another about-face, with the band's frame of reference being drawn from increasingly metallic sources - Treichler himself claims, "we're closer to **Monster Magnet** than **KMFDM**." And TV Sky "is a fucking good rock record." **(see Treponem Pal)**

Discography: The Young Gods (Product 1987); Red Water (Play It Again Sam 1989); L'Eau Rouge EP (Play It Again Sam 1989); Play Kurt Weill (Play It Again Sam 1991); TV Sky (Play It Again Sam 1992); Skin Flowers EP (Play It Again Sam 1992) ; AL COMET LADEP (1992)
EUROP PIRAT TOUR (150 BPM 1992)

INTERVIEW
YOUNG GODS/FRANZ TREICHLER

by Lisa Ridley

You recorded your last album, TV Sky, in English; what prompted the change?
I wrote most of the stuff while I was in New York. I lived in New York last year for two, three months; that's when I actually had the time to gather some material together. So after a month, I started thinking in English, and I wrote most of it in English. Basically all the ideas, even the musical ideas I wrote there. I wanted to do more of a rock LP, an American album.

You wanted to reach an American audience?
A little of both. American audiences like the way things were, they like the cabaret style, the classical European thing. I like French, but French and English are very different to speak. I like both of them for their different styles. English will let me be more melodic; in French, every word is more heavy.

The songs on this record are considerably different to your previous ones.
There was a choice of not sampling anything else, more what a regular band, rock band would be like. No more classically sampled music, no more industrial, whatever. Just a straight rock LP.

How's it going?
Not much in America, but in Europe there is a good response. I've noticed that everywhere we go, there is a mixed crowd. Lots of different kinds of people; metal fans, or Front 242 fans or whatever. Very mixed, and I like it.

Do you plan on touring the States a lot in the future?
It depends. I think we'd actually like to; I think this year was kind of hard, and it's the second time we've toured the U.S. We invested quite a lot of money doing that, and we only have a big turnout in major cities. So we have got to reconsider the fact that maybe next time it would be better to support a band

What do you think of American bands?
I met a few bands, they give me tapes, and I think there is a good new generation of bands, mixing technology and metal, whatever, four, five, six musicians onstage, one sampling, one guitar. I caught a few bands like that.

Your music is really hard to peg down. How would you describe it?
Just put adjectives here! "I think it's wide". I would put lots of contradictory things in. There is tenderness, I would say, for a rock band; I wouldn't say it's industrial or metal. I would say it's today's technology.

As far as this record goes, I think it was something that was there already, and we explored it a bit more. I think if you put all the guitar songs in the first three LPs together, you end up with something pretty close to this. Maybe less angry, maybe more, because of the people. This LP is less Europey, less pompous, because of the classical side. I just realised after the Kurt Weill LP, we had reached the top of using and showing all these European influences, and we should just head somewhere else. You have only a certain amount of time to be creative, and you keep being productive. You have to. The people are going to forget about you if you don't put things out. When you write of your last six or eight months of your life, and you try to share whatever you got from them into music, be creative. I think I personally have been influenced by my surroundings and everything which relates to it. The day to day life, the world, etc.

Do you have anything you'd like to say as far as personal philosophies on your life.
Oh no! I am too young to have a philosophy! I learn every day.

And you just channel it into your music.
If I can. That's the problem if you try to do something, and you can't actually do half of what you would actually like to do. So you just keep on and on.

So it's a struggle to put on vinyl what you hear in your head?
Not a struggle, it's a game. Sometimes you are a good player, sometimes a very bad player.

YOUNG SCIENTISTS see **Ministry**

FRANK ZAPPA/MOTHERS OF INVENTION see **Can, Faust, Punk Rock, Residents**

Z'EV

"I've studied ethnomusicology, but I wouldn't want to describe myself as a master drummer" - Z'ev, 1982

San Francisco based percussionist Z'ev pioneered the use of "found" equipment within his chosen field of instrumentation - since his emergence at the Mabuhay Club in 1978, tubes, bath tubs, kitchenware and springs are thrown, hit and battered to devastating sonic effect.

Included amongst the artists with whom Z'ev has worked are **Boyd Rice**, **Glenn Branca** (the 1982 Symphony #2) and **Throbbing Gristle**. Z'ev himself has also worked under the aliases **Deese**, **Element L**, **GDG**, **Magneet Bond**, **Rax**, **Shaoul**, **Uns**, **Stefan Weisser** and **Yoel**.

Most of Z'ev's recordings are representative; Opus 3, recorded live in a church, has a natural ambience which many of the earlier releases, assembled in conditions of greater atmospheric sterility, lack. **(see Glenn Branca, Boyd Rice, Throbbing Gristle)**

Z'ev

Selected discography: Salts of Heavy Metals (Lust/Unlust 1978); Production and Decay of Spatial Relations (Backstreet 1979); Auf EP (Kremlin 1979); Wipe Out EP (Fetish 1979); Zones (shared LP Subterranean 1980); FFF (Subterranean 1981); # for 3 (shared LP Zensor 1981); Documenta 7 Compilation (Carmen Knobel 1981); Berlin Atonal vols 1 (shared with **Throbbing Gristle** Atonal 1984); Opus 3 (Soleilmoon 1991).

INDUSTRIAL REVOLUTION ADDITIONS by CHASE

Putting together an A-Z book of bands in any style of music is like carjacking. No matter how good you are, you're not always going to find that Jaguar. And on that same note, you might end up with a bucket if you don't do enough homework. In order to remedy this problem, I've compiled a list of some of the other movers and shakers that escaped the initial roundup. One of the reasons why a lot of these bands don't get the recognition they might deserve is that their whole career is based on the other side of the ocean and only a few rare imports made or are making the journey.

A SPLIT SECOND

Like most modern day influential electronic-dance projects, A Split Second's two founding members (**Marc Ickx** & **Chrismar Chayell**) began their collaborating in the early 80's. Although the band has covered a lot of digital domains, their most recognized musical ingredient is their incorporation of ethnic rhythms combined with a classical synth base. While the track "Flesh" was the European groundbreaker for the band (particularly when it was played at 33 RPM instead of at its intended 45 RPM), American DJs gravitated towards the Arabesque floorshaker called "Mambo Witch." A Split Second's popularity eventually took off in the States when they were licensed to Wax Trax and therefore available domestically.

Discography: Flesh 12" (Antler), Ballistic Statues (Ant), Rigor Mortis 12" (Ant), Smell of Buddha 12" (Ant 1987), Scandinavian Bellydance 12" (Ant), Mambo Witch 12" (Ant 1988), From The Inside (Ant), Flesh CDS (Ant), Colosseum Crash 12" (Ant), Another Violent Breed - Live Versions 12" (Ant), Firewalker 12" (Antler-Subway), Backlash 7" (A-S), Kiss Of Fury (A-S 1990), The Parallax View 12" (A-S), Flesh & Fire 1991 Remixes (A-S 1991), Rigor Mortis (91 Remix) 12" (A-S 1991), Flesh ('91 Remix) 12" (A-S 1991).

ATTRITION

Inspired from his Coventry-based fanzine (Alternative Sounds, circa 1980), **Martin Bowes** formed the band Attrition. With influences ranging from **Cabaret Voltaire**, **Joy Division**, and the **Fall**, Attrition originally began as a guitar-bass-drums line-up, which later evolved into a dark electronic creature. Throughout their career, Attrition has been on roughly four labels (Third Mind, Antler, Projekt & Contempo) and with each move their sound has expanded and in someways shift. As a result they've been labeled as everything from a New Beat band to an EBM group to a number of other nondescript terms used to categorize eclectic techno-pop bands. The easiest way to sample Attrition is to locate their CD anthology - available domestically on Projekt - which represents the period between 1984-89. One of the unique hooks that defined early Attrition was the mixture of melodic female singing with Bowes diametrically opposed stark embalmer-like voice... their distinct method of vocal embricating has rarely surfaced in the genre since. CD re-issues of their first three albums will be released through Projekt in North America while an album of atmospheric soundtrack pieces called Ephemera will be released by the German label Artware Audio.

Discography: Deliverance 12" (1984), The Attrition of Reason (Third Mind 1984), The Voice Of God 12" (1984), Shrink Wrap 12" (1985), Smiling, At the Hypogonder Club (Third Mind 1985), In the Realm of the Hungry Ghosts (Third Mind 1986), Take Five EP (1987), Death House (Soundtrack) (1987), At the Fiftieth Gate (Antler 1988), Haydn 12" (Antler 1988), Turn to Gold 12" (Antler 1989), Haydn Remix 12" (Antler 1989), Recollection (Projekt 1990), Thin Red Line 12" (Contempo 1991), A Tricky Business (Contempo 1991), Something in My Eye 12" (1992), The Hidden Agenda (Hyperium 1993), Lip Sync EP (Hyperium 1993)

A;GRUMH

A European duo who have covered the whole gamut from harsh electronic dance music to experimental noisescapes to borderline accessible synth pop. While members of the band have worked on a number of side projects, the least worthy but best known was the studio failure (according to a;GRUMH) called **A Chud Convention**. This noisy collaboration consisted of members of a;GRUMH and **Skinny Puppy**.

BLUE EYED CHRIST

While Belgium has bred a lot of bands that reflect their scene, so has Chicago. When **John Norten**'s Blue Eyed Christ project first surfaced, it mirrored a lot of Chicago's popular in-crowd and consequentially didn't disrupt the torture-tech hierarchy at the time. However, like most bands who begin with a borrowed sound, they either evolve into their own identity or they die in the used bins. Norten's new music still maintains its streetwise edge, but carves a fresh sound out of mixing beat-box influenced grooves and aggressive samples with his jagged-yet-commercially viable vocals. When B.E.C.'s new material hits the street, we'll see if they reach the next plateau or end up pumping gasoline.

Discography: Leaders And Followers (Kk 1991); Catch My Fall EP (remixed with Die Warzau) (Kk 1992).

CAT RAPES DOG

This duo is Sweden's answer to all the growling throngs of **Puppy**-inspired bands elsewhere. **Joel Rydstrom** and **Magnus Fransson**, however, have flushed out their music with something that rarely sees the light in this genre—humor. In an interview in Crewzine, Magnus was cited "We write about serious things and then we twist it. Preferably in an ambiguous way. Some lyrics, though, are just weird and don't mean anything at all." In 1989, **Annelie Bertilsson** enlisted and brought with her a bass guitar that blended well with CRD's rapid bpm and electrocore sound. Since their debut, CRD have been primarily rooted in the more aggressive end of the torture-tech dance spectrum, but they have touched upon melodies as well as less threatening forms of techno.

Discography: Columna Vertebralis EP (Front Music 1988), Maximum Overdrive (Kk 1989), Trans Europa compilation (150 BPM 1989), Fundamental 12" (Kk 1990), Mad Man/True Love 7" (1990), Gods, Guns & Gasoline (Kk 1990), American Dream/Heavenly 7" (1990), Banzai Beats EP (Kk 1991), I Sometimes Wish I Was Famous compilation (Energy 1991), Superliminal 12" (Kk 1991), Trojan Whores 12" (Kk 1993), Moosehair Underwear (Kk 1993)

DIGITAL POODLE

In their six-year existence, **Digital Poodle** have been refining their sound to avoid the typical 'Canadian Industrial' clichés. The DP sound is a melodic hybrid of Euro-EBM and tekno-house, with a penchant for excessive ambient- industrial excursions. Live, the Poodle experience is augmented by two live percussionists, as well as films, slides and general mayhem. Lyrically the content of most DP material stems from vocalist **Mouth 392's** experiences working as a subversive in the Baltic States struggles for independence.

Discography: Pay Attention (Shadow, 1987); Poodle Crematorium (Shadow, 1988); Snivel (Shadow, 1989); Live Death (Shadow, 1990); Baltic Work Force (Shadow, 1991); Soul Crush (Shadow, 1991); Soul Crush (DOVe, 1991); Work Terminal CD (Hyperium, 1992); Elektronik Espionage (DOVe, 1992), Crack 12 (DOVe UK, 1993); Division (forthcoming); Noisea (forthcoming).

DIN

"Din is everywhere!" is the slogan. You car find DIN numbers on aspirins and toilet seats, DIN plugs on your midi gear, and the din of shopping mall techno In all corners of the world. The brainchild of **Digital Poodle** member **Pupka Frey**, the music is influenced heavily by the likes of **Jean-Michel Jarre, Propaganda, Abba** and other pop futurists. Pupka relies upon pure electronics to achieve the warm, seductive atmospheres that have become a trademark of this sometimes misanthropic character.

Discography: Makaroni (1988, Shadow); Makrame (1989, Shadow); Harakiri (1991, Shadow); Fantastic Planet (1992, DOVe); Watersports (1993, DOVe UK); Decade of The Brain (forthcoming).

DIVE

Based around techno icon **Dirk Ivens** (a former member of the **Klinik** and founder of Body records), this project has gained a lot of attention in the European market. Dirk's music combines asphalt beats with a wide array of musical approaches spanning experimental to semi-dance pieces. Dirk also makes an appearance on the **Die Krupps**' song "Dawning of Doom".

Select Discography: Final Report MCD (Minus Habens), First Album (Minus Habens), Broken Meat (Minus Habens), Images MCD (Minus Habens 1993).

FINI TRIBE

Most of America's inundation of Fini Tribe has been either through the two singles on Wax Trax ("Make It Internal" & "I Want More", the latter being a **Can** cover) or by the Tribe's more recent crossover techno fodder. As is often the case with low profile bands, they have a handful of material that predates all of this. Way before **Chris Connelly** was wailing for **Ministry, Pigface, Revolting Cocks** and a bunch of other guest appearance projects, he was fronting the tribal-edged, slightly-discordant rock group the Fini Tribe. Following an extremely rare Peel Sessions cassette release, their "Curling and Stretching" 12" was the first piece of vinyl to expose Connelly's distinct voice. While the Tribe had released material on both their Fini Flex label and Cathexis, it was a remake of "De Testimony" (called "Make it Internal") released on Wax Trax that first hit the club scene in America. Since **Connelly** left the band (and is currently doing his solo projects), the Tribe released some albums with a similar feel, but later degraded into diluted techno, subduing those who like to dance and alienating those who prefer to just listen.

Select Discography: Curling and Stretching 12" (Fini Flex 1984), Let The Tribe Grow 12" (Cathexis), You Bet We've Got Something Against You compilation (Cathexis 1986), I Want More 12" (Wax Trax 1987), Make It Internal 12" (Wax Trax 1987), Noise, Lust & Fun (Fini Flex 1988), Zulus 12" (Fini Flex 1988), Elect-Rolux 12" (Fini Flex 1988), Animal Farm 12" (One Little Indian 1989), Grossing 10K (One Little Indian 1989), Monster in the House 12" (One Little Indian 1990)

FORCE DIMENSION

Back in '85, the Dutch band which began as a trio, quickly slimmed down to its two main members, **Tycho de Groot** and **Rene van Dijck**. After releasing several cassettes and doing some compilations, they were picked up by Kk. Kicking off their signing was the unique dual release of two identical albums, with the exception of one being produced by the band (sporting a red cover) and the other by **Luc Van Acker** (this had a blue cover). This band oddly walks the line between abrasive aggro-tech and commercially viable dance tracks.

Discography: The Force Dimension (Kk 1988), Dust 12" (Kk 1989), Tension 12" (Kk 1989),

Dance The Algorhythm 12" (Kk 1990), Deus Ex Machina (Kk 1990), New Funk 12" (Kk 1992).

INSEKT

When **Mario Varewijck** (ex-**Vomito Negro**) and **Eric Van Wonterghem** (ex-**Klinik**) pooled their talents in the end of the 80's, the result reflected their past associations as well as some new enhancements. On the vocal tip, they've done the fuzzy distortion approach and later evolved into a more coherent style. Musically, the band's current direction still has hard rhythms in addition to late seventies sounds from the Sheffield school.

Discography: We Can't Trust The Insekt (Kk 1989), Control Your Fear...Now 12" (Kk 1990), Stress (Kk 1990), Dreamscape (Kk 1992), Live Creeps (Kk 1993), In The Eye (Kk 1993)

KLINIK

This influential band is like a lot of others in the sense that its children get all the recognition, instead of the parent. The members of the Klinik (**Dirk Ivens, Marc Verhaeghen, Erik van Wonterghem**) went on to form **Insekt, Dive, Noise Unit** and **Blok 57**. Maybe its because these other projects took elements of the Klinik and improved upon it...or maybe America is just slow in the history department. In any case, the Klinik incorporated sinister, straight-from-the-torture-chamber rhythms with crawling vocals and grim aesthetics.

LASSIGUE BENDTHAUS

Originating as a one man project in 1987, L.B. has cultivated some interesting approaches to the electronic field. Although they have a traditional feel with their clean blend of electronics and subdued vocals, their latter day experiments into the science of cloning and how it ties in with music is where they really defined themselves. Their Cloned album focused on the philosophy that nothing is sacred or untouchable, everything can be dissected, re-manipulated, re-organized, re-produced and re-done. In other words, cloned.

Discography: Automotif 12" (Parade Amoureuse/Contempo), Matter (P.A.-Con 1991), Hertz 12" (PA-Con), Biohazard 12" (Contempo 1992), Bio-Feedback 12" (Contempo 1992), Cloned (Contempo 1992), Cloned Binary (Contempo)

LUC VAN ACKER see MUSSOLINI HEADKICK

MALHAVOK

Like other bands in the cybermetal movement, the line becomes blurred on Malhavok's Premeditated Murder release as to whether this Canadian-based project is a metal band with samplers or an electronic band with guitars. The distinct and always present guitar work, however, sways them closer to the headbanger camp. When Metal Blade brought The Release to the U.S., this CD offering combined two albums (The Release & Punishments). Although labelmates **Skrew** attracted all the buzz, Malhavok provides equal promise.

Discography: The Release (Epidemic/Metal Blade), Premeditated Murder (Epidemic/Metal Blade 1993).

MUSSOLINI HEADKICK

Just like it took **Ministry** to open people's ears to the veteran voice of **Chris Connelly**, it took the **Revolting Cocks** to introduce the highly accomplished producer and composer **Luc Van Acker** (Luc not only released his own solo projects, but he participated on **Ministry**'s Twitch album, he produced a **Force Dimension** release, performed with **Anna Domino**, as well as a long list of other worthy credits). In any case, MH teamed up **Acker** with **John Butcher** and **Jean-Pierre Everaerts** to result in a driving collection of electronic-rock tracks. In one of their bios they cite that "At least Mussolini... 1). paid for his fascistic sins, 2). made the trains run on time, and 3). lent his name to a new and exciting band."

Discography: Themes For Violent Retribution (World Domination 1989), Blood On The Flag (World Domination 1990), No Sacrifice 12" (Retribution 1993)

PAIN EMISSION

While L.A. and S.F. have spawned enough electronic-dance bands to fill up a box set of compilations, San Diego has only two bands that have remained consistent in the torture-tech movement. The first is **Sweat Engine** - who released a 12" and appeared on a couple compilation CD's - and the other is Pain Emission. P.E. entered the live electronic domain in 1990 with dark bpm-varied synthcraft, strained vocals and a theatrical approach that dealt with everything from Jonestown reenactments, to flinging animal guts to playing drums with a flaming dead eel. Eventually the novelty of sensationalism was replaced by a desire to present a more traditional image. Now in '93, after having their first full length CD released on the German-based C.O.P. label, the drum machine is being replaced by real percussion and the guitar slinging has become a more predominant element. The first taste of their current direction is available on the Ripped Up... 7" compilation on which the band serves up an exclusive mix as they battle for attention with 16 Volt and two non-electronic rock acts (**Balance** & **Eggbound**).

Discography: The Cyberflesh Conspiracy (IIM 1992), Below San Onofre compilation (Pan Handler Productions 1992), Rivet Head Culture compilation (IIM 1993), Fidget (C.O.P. 1993), Ripped Up & So Sedated (Re-Constriction 1993)

PLACEBO EFFECT

Similar to many other German coldwave bands, Placebo Effect cites **Kraftwerk** and **DAF** as inspiration, but also includes the more experimental **Throbbing Gristle** and **Whitehouse**. Placebo Effect (**Axel Machens, Achim Windel, Christoph Kunze**) combine monumental sound creations of fragile samples, polyrhythmic drum tracks and various vocal distortions to - according to the band - result in soundtrack music for imaginary horror films.

Discography: Techno Club compilation, Rebeat compilation (Noise Zone), Danse Macabre Sampler compilation (Danse Macabre), Galleries of Pain (Danse Macabre 1992)

POESIE NOIRE

Living up to its French name (translated: "black poetry"), Poesie Noire originally juxtaposed dark, somber melodies with hard beats and vocal snarls. After a large number of releases, P.N. started drifting towards more commercial pop ballads and vacated the scene they had become such a staple in. The distinct track, "Pity For The Self", has been one of their best contributions to the U.S. club circuit.

Selected Discography: Tales of Doom (Antler 1987), Timber 12" (Antler), Tetra (Antler), Oblivion 12" (Antler-Subway 1990), En Grande Colere (Antler), Pity For The Self (Antler), Love Is Colder Than Death (Antler)

PORTION CONTROL

Cited as an influence by the likes of **Front 242, Front Line Assembly, Skinny Puppy** and many other aggro-synth bands, this London based trio of **Dean Pavianni, John Whybrew** and **Ian Sharp** started collaborating on studio work back in 1980. One look at their Psycho-Bod LP and there's no mistake about their pioneering role in the cyberpunk world. P.C. is currently reformed under the moniker **Solar Enemy** and have released Dirty vs. Universe on Third Mind and Proceed To Beyond on T.E.Q. Music? (the newly formed label by the fanzine Music From The Empty Quarter).

Selected Discography: I Staggered Mentally (In Phaze), Purge 12" (Dead Man's Curve 1986), Psycho-Bod Saves The World (Dead Man's Curve 1986), Totalbeatfactor compilation 10" (Big Noise In Archgate 1988)

POUPPEE FABRIKK

If you're getting out a translation dictionary, don't bother. The name serves no purpose other than to confuse and bring chaos into the media. Following the current trend of cybermetal, this Swedish former synth project straps on the guitars to distance themselves from the rest of the Energy scene [Energy Records has a large catalogue of aggressive as well as melodic synth pop releases]. While the band gets progressively harsher with each release, they still maintain their electronic-dance edge. To get a feel of the band, check out the Crusader CD which compiles their best music from 1988-1992. On the track "TOTDN", **Joel** (a fellow Swede) from **Cat Rapes Dog** makes a guest vocal appearance.

Select Discography: Summoning MCD (Last Chance), Rage (Energy), Portent (Energy), Betrayal 12", Crusader (Energy 1992)

PROJECT PITCHFORK

When looking for dark electronic bands in America, one must canvass large amounts of territory. In Germany, however, all you have to do is yell really loud. Among the many projects fitting into the cybergoth category, one is often considered a best buy. This is **Peter Stilles**' Project Pitchfork. In addition to releasing his shadowy, hard-edged material, he also has a side project called Aurora.

Select Discography: Entities (Hyperium), Souls Island (Hyperium)

2ND COMMUNICATION

Although Japan's digital threatbeat scene is not large, there does exist two worthy projects in Sapporo. The First is **DRP** (**Tomoyuki** & **Kuniyuki**) whose 1992 debut, Electro Brain 286, appeared on Body Records (**Dirk Iven**'s label) and is being followed up by a release on Kk. The other band is 2nd Communication. 2nd Comm.'s three members (**Dossa Yun, Trast C. Howard, Toshiaki Ishida**) create songs that usually go on for over six minutes with unrelenting throbbing dance beats and white noise vocals. No ballads for these guys. Both bands appeared on Kk's Electro-Genetic compilation (although it didn't showcase either bands' better material) and the Music Hologram 1 compilation released on the Osaka-based Electric Blue Label (a division of Les Disques Du Soleil).

Discography: 2nd Communication (Kk 1989), My Chromosomal Friend (Kk 1990), Brain That Binds Your Body 12" (Kk 1990)

16 VOLT

Formed by **Eric Powell** at the death of the 80's, 16 Volt combines militant sequencing, bloated guitars and brutal throatwork. Even though the guitar and computer formula is not a new combination, 16 Volt veers closer to a garage-influenced guitar sound reminiscent of their Oregon locale, in lieu of the other metal, punk, and thrash approaches. Also, rather than allowing the guitars to dominate, 16 Volt maintains a perfect balance between strings and sequencers. After appearing on an IIM compilation and the soundtrack to The Living End, 16 Volt was signed to Re-Constriction in late '92. The debut CD, entitled Wisdom, featured the production talents of **Keith "Fluffy" Auerbach** and **David Ogilvie**, **Puppy**'s chief engineer.

Discography: The Cyberflesh Conspiracy compilation (If It Moves... 1992), Wisdom (Re-Constriction 1993), Ripped Up & So Sedated 7" compilation (Re-Constriction 1993)

STG

Because the L.A. basin has become a breeding ground for bands who mesh guitars, sequencers, and a vocal processor, the struggle to find an identity means different roads have to be taken. When STG kicked into gear back in 1990, they traveled the road too often taken with the good-yet-commonplace choking on wasp vocal style combined with fat beds of synthwork and sampleplay. After a series of vocalist impeaching, a truckload of cassette releases and a couple compilation appearances, they've finally found their identity. STG - or Screaming To God - takes elements of early 80's raw punk rock - as opposed to the speed metal-&-techno trend - and blends it with their electronic foundation to create a pit provoking hybrid.

Discography: Torture Tech Overdrive compilation (If It Moves... 1990), Rivet Head Culture compilation (IIM 1993), No Longer Human (Rotten Records 1993).

SWAMP TERRORISTS

Many bands have experimented with mixing hip hop and guitars, but few do it as consistently as the snarling two-man team from Berne, the Swiss capitol. Since the band's conception in 1988, they've continued to combine chunky beat-boxing with catchy jabs of metal guitars and heavily accented vocals without bowing down to the market trends. The Swamp Terrorists are probably Switzerland's best kept secret.

Discography: He is Guilty 12" (150 BPM 1989), Trans-Europa Compilation (150 BPM), Grim-Stroke-Disease (Machinery/Noise International 1991), Grow-Speed-Injection (Machinery 1992), Rebuff 12" (Machinery 1992), I Like Texas (the Forbidden DJs - A side project) (Machinery 1992), Pale Torment 12" (coming soon)

VOMITO NEGRO

With a latin name that translates into 'black vomit', it's not difficult to understand why the duo of **Gin Devo** and **Guy Mieghem** have had to work hard since their formation in 1983 to become a staple in the history books of aggro-dance music. Like many bands who have been around for over a half decade, Vomito Negro has encompassed a variety of styles - ranging from aggressive, digitally-biting soundscapes to slow brooding experiments in ambience and noise to full-on dancefloor assaults - as well as hitting upon a variety of topics—the most current one being environmental awareness.

Selected Discography: Vomito Negro (1985), Stay Alive 12" (Kk 1987), Dare (Kk 1988), Dare & More (Kk 1989), Shock (Kk 1989), Save The World 12" (Kk 1989), Human (Kk 1990), The New Drug (1991 Antler-Subway), Wake Up (Antler-Subway 1992).

WILL

Originating back in '87 - yet keeping it underground until the early 90's - Will is the triad of **Chris Peterson, John McRae** and **Front Line Assembly's Rhys Fulber**. Although their heavily orchestrated sound and Biblical art evoke images of a monastery up on a Budapest hillside, their intent is not to preach or to be sacrilegious, but to follow the more traditional route of singing about things they feel spiritually attached to.

Discography: Pearl of Great Price (Third Mind 1991), Word Flesh Stone (Third Mind 1992)

While the list of bands who have contributed to the propagation of this style of music is unlimited - or so it seems to this overworked, non-paid, nonunion contributor - there are still some vital ones that you need to look into to flesh out your knowledge. Therefore, I've made a small list of bands and some of their labels for you to diligently research:

PIG (Concrete, Wax Trax), **SCHNITT ACHT** (Cheetah), **PRAYER TOWER** (Third Mind, Marz), **NON AGRESSION PACT** (GPC Productions), **SA 42** (LD Records), **DAF, SCAPA FLOW** (Parade Amoureuse, Energy), **INSIDE TREATMENT** (Parade Amoureuse, Energy), **SCREAMING TREES UK** (Native), **SNOWY RED** (Antler), **CASSANDRA COMPLEX** (Play It Again Sam), **PERENNIAL DIVIDE** (Sweat Box), **DIE FORM** (Parade Amoureuse, Dead Man's Curve), **CLICK CLICK, HULA, OIL IN THE EYE** (Energy), **TILT!** (Antler-Subway), **NOISE UNIT** (Antler-Subway), **IN SOTTO VOCE, BORIS MIKULIC** (Antler-Subway), **A.A.A.K** (Kk, Scam), **PARANOID** (Machinery), **KODE IV** (Kk), **YEHT MAE** (Zoth Ommog, Gymnastic), **OOMPH!** (Machinery), **SLOPPY WRENCHBODY** (Kk), **SNOG** (Machinery), **BRAINCHILD** (R.E.X.), **DRILL** (Abstract Sounds), **FINAL CUT** (Nettwerk), **VEIN CAGE** (Axis), **SMERSH** (Kk, Dead Man's Curve), **SECOND VOICE** (Hyperium), **MENTALLO & THE FIXER** (Zoth Ommog, Simbiose)...

If you have corrections, additions, or attractions - blatant or minute - to my addendum, please write 'em down and send them to Cleopatra. Strap down. -Chase

A compilation synopsis by Chase

A wiseman once said that if he had a dime for every good torture-tech compilation, he could probably buy something really nice for his ex-wife, although he had no intention of doing so. The wisdom inherent in the heart of this adage rings true with its keen insight regarding the vast number of compilations saturating the market. While we can't focus on every one released, to disregard this category completely would be impaling the fountainhead of the electronic scene. As a result, a select discography of compilations is contained to give the rivet head consumer initial access into the realm of torture-tech (i.e "Industrial"-dance) and noise culture. Unfortunately, the release dates weren't always printed on the products, in addition to the fact that some bands appear only on CD issues.

Bark / Sweat Box (1988)
Although the Sweat Box label is defunct, this compilation showcased the talents of MEAT BEAT MANIFESTO and IN THE NURSERY prior to their Wax Trax exposure.

Body Rapture I & II/ Zoth Ommog (1990)
While kids in the States were busy eating up all the W.Trax and Nettwerk bands, the German-based Zoth Ommog label was busy inundating the overseas market with their own digital-deviants. The Body Rapture compilations featured the likes of BIGOD 20, LEÆTHER STRIP, DIE KRUPPS, INTERMIX, X MARKS THE PEDWALK and the ARMAGEDDON DILDOS.

California Cyber Crash / C.O.P. International (1992)
These new folks on the block have managed to get most of the California torture-tech celebrities onto one release, with everyone from CONSOLIDATED, CONTAGION, and GROTUS to DIATRIBE, KODE IV and XORCIST. This showcase of the new school of torture-tech is a good representation of the West coast scene.

Crack of a Belgian Whip / Kk (1992)
The often overlooked Kk label demonstrates that they're a tough contender in the harsh-electronic dance market by unleashing powerful tracks by CAT RAPES DOG, 2ND COMMUNICATION, FORCE DIMENSION and DOMINION. This double LP also contains moodier and more abrasive cuts by STEREOTAXIC DEVICE, SMERSH, and JASON RAWHEAD, respectively,

Cyberflesh Conspiracy, The / If It Moves... (1992)
Due to the lack of domestic torture-tech labels, I.I.M. continues to a release a compilation each year showcasing a lot of the up and coming or obscure t-tech talent. Among its 16 artists, the Cyberflesh Conspiracy features CHEM LAB, DIATRIBE, BABYLAND, SLEEP CHAMBER, 16 VOLT, PAIN EMISSION, STABBING WESTWARD and the VAMPIRE RODENTS.

Cybernetic Biodread Transmission / Simbiose (1992)
Rather than adhering to geographical boundaries, the Portugal-based Simbiose label picks its beat-oriented entries from a lot of various countries. Highlights include Canada's NUMB, Belgium's DIVE, England's GREATER THAN ONE, U.S.A.'s BIOHAZARD/PCB (now CONTAGION) and Germany's LASSIGUE BENDTHAUS.

Doctor Death's V.III / C'est La Mort (1989)
Unlike the other D.D. compilations, this featured a large assortment of torture-tech merchants (FLA, LEGENDARY PINK DOTS, CONTROLLED BLEEDING, DATA-BANK-A, PSYCHE) melded with a handful of 4-ADish atmospheric projects.

Dry Lungs I-III/ Placebo
Noise to ambient soundtrack music. CRANIOCLAST, G.T.O., DIE FORM, ESPLENDOR GEOMETRICO, SLEEP CHAMBER...

Ears To The Grindstone / K.O. City Studio (1987)
One of America's torture-tech compilation pioneers is the no longer existing K.O. City Studio label. Their four compilations - Back to the Grindstone (1988), Turn of the Grindstone (1989), Endless Grindstone (1990) - cover the whole electronic spectrum from ambience to noise to dance. The less obscure participants were DATA-BANK-A, COMPOUND, DOMINION and CARBON 14.

Electronic Body Music, Another World Pt.III / Antler-Subway

Keeping in a hard-edged torture-tech vein, this release has a respectable line-up including A SPLIT SECOND, VOMITO NEGRO, OVERLORDS, KLINIK, MUSSILINI HEADKICK, CODE INDUSTRY, IN SOTTO VOCE...

Fight! / Cathexis (1987)
This release was definitely ahead of its time with its line-up of current stars. The REVOLTING COCKS unleash a raw rendition of T.V. Mind, while THE YOUNG GODS, CLICK CLICK, IN THE NURSERY, HULA and SOVIET FRANCE earn their stripes as well.

Funky Alternatives vol.I-VII / Concrete Productions
Launched in 1986, the FUNKY ALTERNATIVES series went out and signed the most relevant and indepedent artists making left field dance music such as NEW ORDER, TACKHEAD and DAF. The sequels have continually kept up to date with the scene including the likes of CABARET VOLTAIRE, KMFDM, NITZER EBB, SKINNY PUPPY, FRONT LINE ASSEMBLY, THRILL KILL KULT, G.T.O., CLICK CLICK and many more.

How To Use Machinery / Machinery (1992)
Following in the abrasive-dancehall digital terrorism of Zoth Ommog, Kk and Wax Trax, Machinery introduces a powerful new rogues gallery of floorslammers including SWAMP TERRORISTS, SNOG, OOMPH!, DANCE OR DIE and AND ONE.

Hurt / Braindrop (1991)
Centered around a West coast radio show on KXLU (Psychotechnics), this release serves up some raw dance-to-noise material from the likes of BABYLAND, STEREOTAXIC DEVICE, YEHT MAE, and GEKO to other obscure digital-urchins from the LA basin.

If You Can't Please Yourself, You Can't Please Your Soul / SOME BIZARRE (1985)
This compilation melded together some of the more "industrial" trend setters with the torture-tech movement. Notables include EINSTURZENDE NEUBAUTEN, TEST DEPT., YELLO, SCRAPING FOETUS OFF THE WHEEL, PSYCHIC TV, COIL, and CABARET VOLTAIRE.

I Sometimes Wish I Was Famous / Energy (1991)
Taking a more entertaining approach is this Swedish compilation of DEPECHE MODE cover songs. The MODE did some major paving on the more popular roads with their mix of sampled industrial machinery and accessible rhythms, so it's no surprise that aggressive bands like CAT RAPES DOG, POUPPEE FABRIKK and INSIDE TREATMENT pay tribute. Most of this release, however, falls into the vein of synth-pop (over torture-tech) and these softer tracks consequently don't beat out the tried-and-true MODE.

Latex TV Oblivion / Minus Habens (1992)
This Italian label demonstrates that they can hold their own in the compilation coldwar. Contributors include FRONTLINE ASSEMBLY, BATZ WITHOUT FLESH, LASSIGUE BENDTHAUS, DOMINION, BLACKHOUSE, DIVE...

Life At The Top / Third Mind (1984)
LEGENDARY PINK DOTS, ATTRITION, COIL, BUSHIDO, MUSLIMGAUZE... Some other T.M. compilations include Rising From The Red Sand (1983), For Your Ears Only (1987), Une Saison En Enfer (1988), The Third Mind (1990), Mindfield (1992).

Metro Tekno / Silent (1992)
While Canada is part of America, little is known about its torture-tech community other than the fact that SKINNY PUPPY, FLA and NUMB all have origins there. Fortunately, the Shadow label compiled this line-up of Canadian-based t-tech projects to open the aural doors. DIGITAL POODLE, MONA LISA MESCALINE and DIN are three of the more popular acts.

Mindfield / Third Mind (1992)
FLA, CONTROLLED BLEEDING, WILL, SOLAR ENEMY, DELERIUM, EDWARD KA-SPEL...

1992 / Kk (1989)
Before Kk's Crack of A Belgian Whip, there was 1992. This release combined some experimental pieces with some more digestible dancefloor edibles. Highlights include FORCE DIMENSION, VOMITO NEGRO and H.N.A.S..

Perdurabo / Cathexis/Black Rose (1987)
This is a diverse compilation, but some of the points of interest include ATTRITION, LEGENDARY PINK DOTS, WOLFGANG PRESS...

Rivet Head Culture / If It Moves... (1993)
Continuing in the IIM tradition, a bagful of new or neglected harsh electronic dance artists are showcased on this 19 band CD. Highlights include RAW DOG (featuring Ogre, Rose & David Ogilvie), CHEM LAB, OUT OUT, NON-AGGRESSION PACT, and RECLINER.

Somewhere in the Skeleton / Body Records (1990)
Compiled by Dirk Ivens (KLINIK), this release features some obscure but quality torture-tech acts like TILT!, PARANOID and the asian DRP..

Technoclub V.I & II / Music Research
While the first disc was fair with its good line-up (FORCE DIMENSION, SA 42, INSIDE TREATMENT...), the follow-up double CD did a better job. While one half is dedicated to straight techno, the other half is growl-styled torture-tech with SKINNY PUPPY, PANKOW, MINISTRY, BLOCK 57, FLA, PROJECT PITCHFORK, PLACEBO EFFECT...

Technopolis V.I-III / Antler / Music Research
Because M.R. is the blanket label for Zoth Ommog and a few others, it's no surprise that the line-up is so strong. Although the first volume contained the likes of CONSOLIDATED (while the rest were more obscure acts), it wasn't until Vol.III that they really opened eyes with LEÆTHER STRIP, ARMAGEDDON DILDOS, X MARKS THE PEDWALK, KLUTE, A-HEAD...

The Dice are Rolling / Play It Again Sam
This release showcased the bigger names from the PIAS catalogue like a;GRUMH, LEGENDARY PINK DOTS, THE WEATHERMEN, SIGLO XX and the NEON JUDGEMENT to name a few.

The Elephant Table Album / X Tract 1983/1989
Originally compiled in 1983, this album was repackaged in 1989 containing 20 of the 21 original tracks. It's a double album with most of the tracks focusing on experimental noise, but the dance-oriented bands (PORTION CONTROL, 400 BLOWS, NOCTURNAL EMISSIONS) are the ones that most new school t-tech projects cite as influences. Other acts include CHRIS & COSEY, COIL, NURSE WITH WOUND, MUSLIMGAUZE, ATTRITION, PINK DOTS...

The Last Testament / Fetish
A hard to acquire compilation featuring a lot of the old-school. CLOCK DVA, THROBBING GRISTLE, 23 SKIDOO, Z'EV...

The Third Mind / Third Mind 1990
This is almost a brother/sister release to the Mindfield compilation. DELERIUM, FLA, SOLAR ENEMY, IN THE NURSERY, SON OF SAM...

This is Electronic Body Music / PIAS, RRE, Antler & Nettwerk 1988
Combining the best of four labels results in this hard-edged club-happy behemoth. Every band listed has a strong history, from FRONT 242, CLICK CLICK, and CASSANDRA COMPLEX, to A SPLIT SECOND, BORGHESIA and SKINNY PUPPY.

Torture Tech Overdrive / If It Moves... 1991
This predecessor to the CYBERFLESH CONSPIRACY was available on vinyl-only and limited to a 1000 copies. In addition to coining the now common literary term "torture tech" (to be used in lieu of the misnomer"'industrial'-dance"), this release exposed the likes of CONTAGION, XORCIST, YEHT MAE and A-POLITIQ;.

Total / Parade Amoureuse 1991
A lot of old-school torture-tech and borderline noise-culturists fill the disc, from FRONT LINE ASSEMBLY, FINI TRIBE, and BOURBONESE QUALK to COIL, THE ANTI GROUP and FLUX.

TotalBeatFactor / Big Noise in Archgate 1988
Although it was released as a 10", it still packed an impressive line-up consisting of POESIE NOIRE, FRONTLINE ASSEMBLY, PORTION CONTROL and three others.

Trans Europa / 150 BPM 1989
Utilizing a duel-country format (Swiss bands on one side, Swedish on the other), they managed to capture the best t-tech from each territory. Highlights include SWAMP TERRORISTS, CAT RAPES DOG, and SCAPA FLOW.

Tyranny of the Beat / Mute (1992)
This release definitely packs a historical edge with SPK, THROBBING GRISTLE, CABARET VOLTAIRE, DAF, DIE KRUPPS, EINSTURZENDE NEUBAUTEN, CAN, NON and plenty more to fill your blender.

Virtual Energy / Energy (1992)
A nice array of non-American torture-tech projects that will hopefully break into the domestic market. OIL IN THE EYE, INSIDE TREATMENT, POUPPEE FABRIKK, SCAPA FLOW...

World Of Electronic Body Music / Antler
In the tradition of ANTLER music, this compilation serves up their full charcoal-beat dance spectrum from the KLINIK, ATTRITION, SNOWY RED, and A SPLIT SECOND, to COMPANY OF STATE, NEON JUDGEMENT and IN SOTTO VOCE.